D1130612

ELECTRONIC
COMPONENTS HANDBOOK
FOR CIRCUIT DESIGNERS
BY R. H. WARRING

TAB BOOKS Inc.
BLUE RIDGE SUMMIT, PA. 17214

FIRST EDITION

FIRST PRINTING

Copyright © 1983 by TAB BOOKS Inc.

Printed in the United States of America

Library of Congress Cataloging in Publication Data

Warring, R.H. (Ronald Horace), 1920-
    Electronic components handbook for circuit designers.

    Includes index.
    1. Electronic circuit design. I. Title.
TK7867.W285   1983   621.3815′3   82-19406
ISBN 0-8306-0493-6
ISBN 0-8306-1493-1 (pbk.)

Cover photograph by the Ziegler Photography Studio of Waynesboro, PA.

Contents

Introduction

Too many books on electronics assume that you already have a profound knowledge of circuits and circuit design. They give circuits based on specific components, and you are stuck with a particular type of integrated circuit or number of transistors and have to match values of resistors, capacitors, etc. You will not necessarily learn very much about electronics if you build such circuits, which may or may not work the first time. Even worse, you could find some of the specified components not readily obtainable. This means you would desperately have to seek advice on suitable substitutes from some expert. And if that expert is not familiar with that particular type of circuit, he may equally be stuck on the same problem or advise you to start again with an alternative circuit. You could have wasted quite a few dollars on components that are no longer going to be of any use to you—especially if you do not understand how electronic circuits work.

It would be much better if you could help yourself in such cases, and that is what this book is really about. Not a collection of circuits to build—but the basis of designing such circuits and understanding how they work. This means starting with an understanding of the components and their basic working principles, and a full appreciation of their practical properties, values, etc. It is the components, put together in matching array, which makes the circuits work. Hence the title of this book. But this is only the beginning of the story. This book also covers the working principles when using components, how to work out required values, etc., and it describes in detail the basis of many standard circuit, designs and the necessary calculations needed to arrive at the suitable component values to use. Designing transistor circuits to match the characteristics of a given transistor, for example, is quite easy when you know how. You will find this explained in some detail.

When it comes to integrated circuits, the problems are a little different. Some, like op amps, are easy to understand and use. Most of the basic op amp circuits are shown, but there are literally hundreds of other possibilities where the component values used with the IC are specifically tied to the individual op amp used. Even more so is the case of the more sophisticated ICs which contain complete circuits or sub-circuits. In this case descriptions must specifically relate to an individual IC. You will find this treatment adopted in describing analog switches, logic gates, electronic audio and radio controls, and rhythm generators.

Hopefully the book covers most of the subjects you would like to know (or need to know) to make electronic circuits an understandable subject. It should also be a reference for information you may need at various times, as well as a design guide with examples that are easy to follow.

Chapter 1

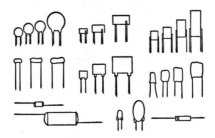

Resistors (Practical)

Resistors are the main elements used in circuit design to arrive at the required current and voltage values, required by the circuit. They work in just the same way in both dc and ac circuits, their performance is not affected by the frequency of an ac supply (although there are some exceptions to this general rule). The parameters we are concerned with in the use and selection of resistors are:

☐ *resistance value* in ohms.
☐ *tolerance* or the possible variation in actual resistance value from the specified value.
☐ *power rating* in watts.
☐ construction or type of resistor.

RESISTANCE VALUES

The range of resistance values to which resistors are made is based on steps giving an approximately constant *percentage* change in resistance from one value to the next, i.e., 1, 1.2, 1.5, 1.8 etc—not simple arithmetical steps like 1, 2, 3 etc. These are known as preferred numbers or *preferred values*. They represent the practical values of resistors obtainable. Here is a list grouped in four ranges—under 10 ohms, 10 - 820 ohms, then kilohms (k ohms) and megohms (M ohms).

Whatever resistor value is calculated for a circuit design the value actually used in the circuit will have to be the nearest *preferred*

under 10 ohms	ohms	k ohms		M ohms
0.33	10	1.0	100.0	0.27
0.5	12	1.2	120.0	0.33
1	15	1.5	150.0	0.39
1.5	18	1.8	180.0	0.47
2	22	2.2	220.0	0.56
3	27	2.7		0.68
3.3	33	3.3		0.82
3.9	39	3.9		1.0
4	47	4.7		1.2
4.7	56	5.6		1.5
5	68	6.8		1.8
5.6	82	8.2		2.2
6	100	10.0		2.7
6.5	120	12.0		3.3
8	150	15.0		3.9
	180	18.0		4.7
	220	22.0		5.6
	270	27.0		6.5
	330	33.0		8.2
	390	39.0		10.0
	470	47.0		12.0
	560	56.0		15.0
	680	68.0		18.0
	820	82.0		22.0

value. Specific calculated values are just not obtainable—unless they happen to work out as a preferred value. Necessary adjustment of calculated values in this way is unlikely to have any significant effect on simple circuit design performance. Quoted values for resistors are, in fact, only nominal and actual values may vary by up to 20 percent either way (i.e. have a *tolerance* of plus or minus 20 percent).

However, if the circuit design is critical, resistors with a much closer tolerance can be chosen—e.g., a tolerance of only 2 percent or even 1 percent. And if the circuit is that critical, then *preferred* (i.e. obtainable) values should be used in the design calculations. This is quite simple to do. Make preliminary calculations to determine the theoretical resistor value(s) required. Adjust these to preferred values. Then re-calculate the circuit performance. If necessary, repeat with alternative preferred values to arrive at an acceptable solution.

COLOR CODES AND TOLERANCES

Regardless of size, shape and type, the usual method of marking a resistor *value* is by colored bands on the body, read in order 1,

2, 3 from the band nearest one end (Fig. 1-1). The equivalent value is determined by:

Color	1 gives first figure of resistance value	2 gives second figure of resistance value	3 gives number of zeros to put after first two figures
Black	0	0	None
Brown	1	1	0
Red	2	2	00
Orange	3	3	000
Yellow	4	4	0000
Green	5	5	00000
Blue	6	6	000000
Violet	7	7	0000000
Grey	8	8	00000000
White	9	9	000000000

Example: resistor color code read as *brown, blue, orange.* The value is read as: brown equals 1, blue equals 6, and orange equals 000. So this register equals 16,000 Ω or 16k Ω (kilohms).

On some old-type resistors the three colors are applied in a different manner—a colored body, colored tip and colored spot on the body. The code is then read in the order: body, tip, spot. Many resistors have an additional (fourth) colored band, which expresses the *tolerance* on the resistor value. The equivalent tolerances so indicated are:

Silver band	10 percent tolerance
Gold band	5 percent tolerance,
Red band	2 percent tolerance
Brown band	1 percent tolerance

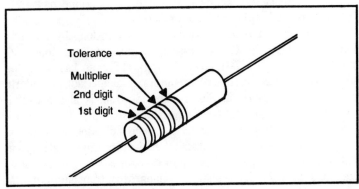

Fig. 1-1. Resistor color coding.

For most circuits resistors with a gold band (5 percent tolerance) are the recommended choice; silver band (10 percent tolerance) are good enough for general use. Finer tolerances (red or brown band) are not necessary, except for critical circuits. No fourth band implies a tolerance of 20 percent. Other variations which may be found on color coding are: *a salmon pink* band indicating a high-stability type, and *a double ring* of the first color band, indicating a wire-wound resistor.

Some types of modern resistors are not color coded but the values are marked directly on to them in numbers and letters. *Numbers* indicate numerical values and *letters* indicate multipliers, viz:

$$R = \times 1$$
$$k = \times 1,000$$
$$M = \times 1,000,000$$

Tolerances are also given by a second letter, viz:

$$M = 20 \text{ percent}$$
$$K = 10 \text{ percent}$$
$$J = 5 \text{ percent}$$
$$H = 2.5 \text{ percent}$$
$$G = 2 \text{ percent}$$
$$F = 1 \text{ percent}$$

The first letter ('multiplier') comes in a position which determines the decimal point. Thus:

rating 5R   designates $5 \times 1 = 5$ ohms
    4k7 designates $4 \times 1,000$ and $0.7 \times 1000 = 4.7$k ohms
    82k designates $82 \times 1,000$ or 82k ohms

The tolerance letter comes at the end, e.g., 4k7J means a 4.7k ohm resistor with a 5 percent tolerance.

Actual *sizes* of resistors range from 4 mm long by 1.2 mm diameter (with a 1/20-W power rating); up to about 50 mm long by 6 mm diameter (with a 7-W power rating); and even larger in the old-fashioned type of carbon rod resistors. A comparison of modern resistor sizes is given in Fig. 1-2 this diagram is also a useful method of identifying the various *types* of resistors by their shape. Note that while the *size* of resistor is a rough indication of its power rating (the smaller the physical size the lower the power rating as a general rule), size is no indication of resistor *value*. Tiny resistor sizes can have very high resistance values. Most transistor circuits operate with low voltages and thus low power ratings, so the smallest sizes of transistors are normally used.

4

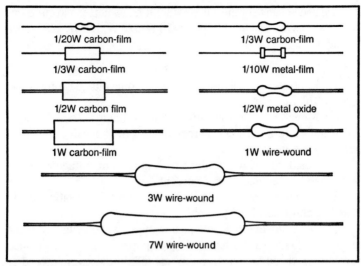

Fig. 1-2. Types of resistors approximate actual size.

VOLTAGE RATING

Maximum operating voltage may also be specified for resistors, but since this is usually on the order of 220 volts or more, this parameter is not important when choosing resistors for battery circuits. Resistors used on electrical mains circuits must, however, have a suitable voltage rating.

TYPES OF RESISTORS (CONSTRUCTION)

There are five main *types* of resistors, classified by their construction.

Molded-Carbon Type

These are literally a rod made from a mixture of carbon and binder fired into a rigid form. The rod is usually protected with a lacquer coating, or a paper or ceramic sleeve. These are the old-fashioned type which are still available (usually as 'surplus' stock) and are suitable for any non-critical circuit. Ratings are usually ¼ W, ½ W and 1 W. Values for this type of resistor range from 10 ohms to 22 M ohms.

High-Stability Carbon Resistors

These are much smaller and made from carbon film. These are miniature and sub-miniature in size and are the preferred type for general use because of their good stability, low noise, and quite low cost. Ratings are usually 1/20 W, 1/10 W, 1/8 W, 1/3 W, 1/2 W, 3/4 W and 1 W.

The most commonly used types are:

☐ Micro—⅛ W usually with a maximum voltage rating of 150 V and a tolerance of ± 5 percent.

☐ Miniature—⅓ W usually with a maximum voltage rating of 250 V and a tolerance of ± 5 percent up to 1 M ohm value; and ± 10 percent for higher values.

☐ Standard—½ W usually with a maximum voltage rating of 500 V and a tolerance of ± 5 percent.

☐ Power—1 W usually with a maximum voltage rating of 750 V and a tolerance of ±5 percent.

☐ High power—3 W metal film type, wire wound type for lower resistance values and/or higher power ratings. Resistance values with these types are often marked on the body instead of color coded.

Since these are the most commonly used types, the values *normally* available in each size are given below.

⅛ W micro—size approx. 4 mm long × 1.8 mm diameter.

	10	100	1k	10k	100k
	15	150	1.5k	15k	150k
2.2	22	220	2.2k	22k	220k
3.3	33	330	3.3k	33k	330k
4.7	47	470	4.7k	47k	470k
6.8	68	680	6.8k	68k	680k

⅓ W miniature—size approx. 8 mm long × 2.8 mm diameter.

1	10	100	1k	10k	100k	1M	10M
1.2	12	120	1.2k	12k	120k	1.2M	
1.5	15	150	1.5k	15k	150k	1.5M	
1.8	18	180	1.8k	18k	180k	1.8M	
2.2	22	220	2.2k	22k	220k	2.2M	
2.7	27	270	2.7k	27k	270k	2.7M	
3.3	33	330	3.3k	33k	330k	3.3M	
3.9	39	390	3.9k	39k	390k	3.9M	
4.7	47	470	4.7k	47k	470k	4.7M	
5.6	56	560	5.6K	56k	560k	5.6M	
6.8	68	680	6.8k	68k	680k	6.8M	
8.2	82	820	8.2k	82k	820k	8.2M	

½ W standard—size approx. 10 mm long × 3.8 mm diameter.

1	10	100	1k	10k	100k	1M	10M
	11	110	1.1k	11k	110k		
1.2	12	120	1.2k	12k	120k	1.2M	
	13	130	1.3k	13k	130k		

6

1.5	15	150	1.5k	15k	150k	1.5M
	16	160	1.6k	16k	160k	
1.8	18	180	1.8k	18k	180k	1.8M
	20	200	2k	20k	200k	
2.2	22	220	2.2k	22k	220k	2.2M
	24	240	2.4k	24k	240k	
2.7	27	270	2.7k	27k	270k	2.7M
	30	300	3k	30k	300k	
3.3	33	330	3.3k	33k	330k	3.3M
	36	360	3.6k	36k	360k	
3.9	39	390	3.9k	39k	390k	3.9M
	43	430	4.3k	43k	430k	
4.7	47	470	4.7k	47k	470k	4.7M
	51	510	5.1k	51k	510k	
5.6	56	560	5.6k	56k	560k	5.6M
	62	620	6.2k	62k	620k	
6.8	68	680	6.8k	68k	680k	6.8M
	75	750	7.5k	75k	750k	
8.25	82	820	8.2k	82k	820k	8.2M
	91	910	9.1k	91k	910k	

1W power—size approx. 16 mm long × 6.8 mm diameter

10	100	1k	10k	100k	1M	10M
12	120	1.2k	12k	120k	1.2M	
15	150	1.5k	15k	150k	1.5M	
18	180	1.8k	18k	180k	1.8M	
22	220	2.2k	22k	220k	2.2M	
27	270	2.7k	27k	270k	2.7M	
33	330	3.3k	33k	330k	3.3M	
39	390	3.9k	39k	390k	3.9M	
47	470	4.7k	47k	470k	4.7M	
56	560	5.6K	56k	560k	5.6M	
68	680	6.8k	68k	680k	6.8M	
82	820	8.2k	82k	820k	8.2M	

3 W 'high power"—wirewound type (typical size 10.5 mm long × 6 mm diameter for resistance values below 10 ohms; and metal film type (typical size 17 mm long × 5 mm diameter) for resistance values from 10 ohm to 22 k ohms.

Metal-Film Resistors

These are made by depositing a film of nickel-chromium on a high-grade ceramic body and then cutting a helical track through the film to produce the required resistance values. End caps are fitted to carry the leads and the resistor body is protected by a lacquer coating. These are even more stable than carbon resistors but cost about four times as much. Ratings are from 1/10 W upwards. The most used type are thick film resistors with a ½W power rating and a tolerance of ± 1% and maximum working voltage of 200 V. Typical size approx. 7 mm long = 2.5 mm diameter. Typical values available are:

1 ohm	10	100	1k	10k	100k	1M
		110	1.1k	11k	110k	
	12	120	1.2k	12k	120k	
		130	1.3k	13k	130k	
	15	150	1.5k	15k	150k	
		160	1.6k	16k	160k	
	18	180	1.8k	18k	180k	
	20	200	2k	20k	200k	
2.2 ohm	22	220	2.2k	22k	220k	
	24	240	2.4k	24k	240k	
	27	270	2.7k	27k	270k	
	30	300	3k	30k	300k	
	33	330	3.3k	33k	330k	
	36	360	3.6k	36k	360k	
	39	390	3.9k	39k	390k	
	43	430	4.3k	43k	430k	
4.7 ohm	47	470	4.7k	47k	470k	
	51	510	5.1k	51k		
	56	560	5.6k	56k		
	62	620	6.2k	62k		
	68	680	6.8k	68k		
	75	750	7..5k	75k		
	82	820	8.2k	82k		
	91	910	9.1k	91k		

Metal-Oxide Resistors

These are made by depositing a film of tin oxide on a special glass rod, the whole being subsequently covered with a heat resistant coating. Stability is again very high and this type is virtually proof against damage through overheating (e.g. when soldering) and is also unaffected by dampness. The usual rating is ½ W with a tolerance of ± 2 percent. Typical size approx. 10 mm long × 3.8 mm diameter. Typical values available are:

10	100	1k	10k	100k	1M
11	110	1.1k	11k	110k	
12	120	1.2k	12k	120k	
13	130	1.3k	13k	130k	
15	150	1.5k	15k	150k	
16	160	1.6k	16k	160k	
18	180	1.8k	18k	180k	
20	200	2k	20k	200k	
22	220	2.2k	22k	220k	
24	240	2.4k	24k	240	
27	270	2.7k	27k	270k	
30	300	3k	30k	300k	
33	330	3.3k	33k	330k	
36	360	3.6k	36k	360k	
39	390	3.9k	39k	390k	
43	430	4.3k	43k	430k	
47	470	4.7k	47k	470k	
51	510	5.1k	51k	510k	
56	560	5.6	56k	560k	

62	620	6.2k	62k	620k
68	680	6.8k	68k	680k
75	750	7.5k	75k	750k
82	820	8.2k	82k	820k
91	910	9.1k	91k	910k

Wire-Wound Types

These are generally only required for special circuits where very low resistance values are required and/or very high currents have to be carried. Typical ratings may change from 1 to 5 W for a 0.5 ohm resistor to 25 W or more for higher resistance values.

EFFECT OF TEMPERATURE

All components carrying a current produce some loss of electrical energy which is turned into heat, causing a rise in temperature of that component. The greater the resistance offered by the component, the greater the heating effect and thus likely temperature rise. It is also a characteristic of most materials capable of passing an electric current that their resistance increases with increasing temperature, i.e. have a positive *temperature coefficient*. One would therefore expect (and it is often quoted) that the effective value of a resistor will increase when actually working in the circuit, due to its rise in temperature. However, this is not a general rule. Carbon is the exception in that it has a *negative* temperature coefficient (*decreasing* resistance with increasing temperature). Whether resistors actually increase or decrease in resistance with increasing temperature depends on the type of resistor involved.

Carbon film resistors have a *negative* temperature coefficient, the value of which will depend on size (wattage rating) and resistor value. Typical figures are:

□ ⅛ W micro: −300 ppm/deg.C up to 100 k ohms in value rising to
−500 ppm/deg.C at 1 M ohm.

□ ⅓ W mini: −250 ppm/deg.C up to 10 k ohm in value rising to
−1000 ppm/deg.C at 10 M ohm.

□ ½ W upwards: typically −180 to −500 ppm/deg. C for all values.

By contrast both metal-film and metal-oxide resistors have a *positive* temperature coefficient, but of small order (which accounts for their superior stability). Typical values are:

□ Metal film resistors: less than + 100 ppm/deg.C.

☐ Metal oxide resistors: typically + 60 ppm/deg.C.
Note: ppm means parts per million.

EFFECT OF AGE

All resistors can be expected to undergo a change in resistance with age. This is most marked in the case of carbon-composition resistors where the change may be as much as 20 percent in a year or two of use. In the case of carbon-film and metallic-film resistors, the change is seldom likely to be more than a few percent.

EFFECT OF HIGH FREQUENCIES

The general effect of increasing frequency in *ac* circuits is to decrease the apparent value of the resistor, and the higher the resistor value the greater this change is likely to be. This effect is most marked with carbon-composition and wire-wound resistors. Carbon-film and metal-film resistors all have stable high frequency characteristics.

VARIABLE RESISTORS (POTENTIOMETERS)

Variable resistors, or potentiometers, are normally constructed from a resistive element, formed into a 270° arc, with a wiper arm connected to it, and turned by a central spindle. There are three external terminal tags; the outer ones connect to the ends of the resistance element and the center tag to the wiper (Fig. 1-3).

The resistance element may be a carbon track or a winding of resistance wire (wire-wound potentiometers). Carbon-track potentiometers are cheaper than wire-wound types and are suitable for most general circuit applications, at low power levels, e.g. ¼ to

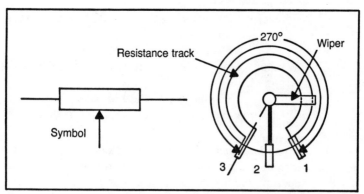

Fig. 1-3. Variable resistor or potentiometer.

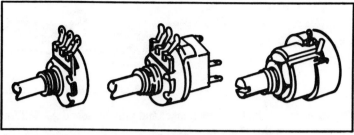

Fig. 1-4. Examples of potentiometers, wire wound type on right.

⅓ W for low resistance values, reducing the power with higher resistance values. Wire-wound potentiometers have a higher power rating—usually of the order of 1 to 3 W continuous for the whole track; they are also made in lower resistance values obtainable with carbon-track potentiometers (see also Fig. 1-4).

Resistance values usually available in each type are:

☐ *Carbon-track*: 100, 220, 470, 1k, 2.2k, 4.7k, 10k, 22k, 47k, 100k, 220k, 470k, 1M, 2.2M and 4.7M.

☐ Wire-wound: 10, 22, 47, 100, 220, 470, 1k, 2.2k, 4.7k, 10k, 22k, and 47k.

Tolerances on potentiometers are usually of the order of ± 10 percent to ± 20 percent, but can be lower with *precision potentiometers.*

Wiring connections to a potentiometer are simple, remembering that the two outside tags connect to each end of the resistance track and the center tag to the wiper. For a simple volume control, connection is thus normally to one end tag and the center tag. Some potentiometers are made with an 'off' location at one end of the wiper movement (usually maximum counter-clockwise movement) when the wiper runs off the resistance track. This type can combine the duties of volume control and on-off switch, in suitable circuits. Power supplies are usually switched on and off by a switch mechanically coupled to the volume control.

The usual connection of potentiometers is 'right-handed' for counter-clockwise rotation to progressively *decrease* resistance in the circuit, e.g., used as a volume control. Specifically, in this case connection would be made to tags 1 and 2. The opposite mode of working—i.e. increasing resistance in the circuit with clockwise rotation requires connections to tags 2 and 3.

Slider-type potentiometers (Fig. 1-5) have a straight track instead of a circular track, otherwise the operating principles are the same. The resistance track may be carbon or wire-wound. Such

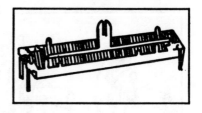

Fig. 1-5. Slider-type potentiometer.

designs can be made very compact and fully enclosed, except for a slot in which the slider arm travels.

Potentiometers (pots) are described as linear or non-linear, this referring to the manner in which resistance values vary with wiper movement and *not* whether the potentiometer is a slider or rotary-action type. With a *linear* potentiometer the resistance change is directly proportional to the actual movement, e.g., a 50 percent movement will correspond to a 50 percent change in resistance (see Fig. 1-6). Slide-type potentiometers normally automatically provide linear characteristics. Rotary types may have linear or logarithmic characteristics, the latter producing an increasingly greater change in resistance with spindle rotation.

Potentiometers may also be designed with characteristics intermediate between linear and logarithmic, e.g., semi-log or linear-tapered; and also with inverse characteristics, e.g., anti-log. Either linear or logarithmic potentiometers are suitable for most volume control duties etc., although logarithmic types are generally preferred.

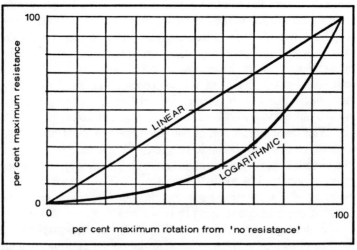

Fig. 1-6. Potentiometer characteristics.

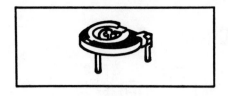

Fig. 1-7. Preset potentiometer

There is also a class of variable resistors intended to be adjusted to a particular resistance setting and then left undisturbed. These are known as *preset potentiometers* (alternatively, *preset pots* or just *presets*). They are small in size and more limited in maximum resistance value—typically from 100 ohms to 1 megohm. They are usually designed for adjustment by a screwdriver applied to the central screw, or sometimes by a knurled disc attached to the central spindle, carrying the wiper. The latter type are known as *edge presets* and similar types, with a (maximum) resistance of 5k ohms may be used as volume controls on miniature transistor radios (see Fig. 1-7).

Chapter 2

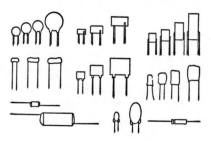

Resistors (Theoretical)

The theoretical resistance of any conductor is given by the formula:

$$R = \frac{\rho L}{A}$$

where ρ is the *specific resistance* or *resistivity* of the conductor material
L is the length of conductor
A is the cross sectional area of the conductor

Conductors are made from materials with a low specific resistance. Resistors are made from materials with an extremely high resistivity, the resulting resistance for a given size of L and A being so enormous that virtually no current can flow through them.

Examples of the use of the basic formula:

☐ A length of wire has a resistance of 4.2 ohms. What is the resistance of the same wire size but four times the length?
Ans: The values of ρ and A are the same, so for four times the length:

$$R = 4 \times 4.2 = 16.8 \text{ ohms}$$

☐ A length of wire has a resistance of 2.8 ohms. What is the resistance of a length of wire of the same material, but twice as long as one half the cross sectional area?

$$\text{Ans: } R = 2.8 \times \frac{2}{\frac{1}{2}}$$
$$= 11.2 \text{ ohms}$$

☐ Calculate the resistance of 1,000 yards of copper wire 0.02 in diameter, given that the resistivity of copper is 0.6 microhm per cubic inch.

Ans: (Express all values in consistent units first)

$\rho = 0.6 \times 10^{-6}$ ohms per cubic inch

$L = 1,000 \times 36$ inches

$A = .7854 \times (0.02)^2$ square inches

then $R = \dfrac{0.6 \times 1,000 \times 36}{10^6 \times .7854 \times (0.02)^2}$

$= 76.3865$ ohms

RESISTORS IN SERIES

Where two or more resistors are connected in *series* their *total* resistance is the sum of their individual resistances (Fig. 2-1):

$$R_T = R1 + R2 + R3 + ...$$

Example: What is the total resistance when resistors of values 220 ohms, 470 ohms and 1.2 kilohms are connected in series.

Ans. $R_T = 200 + 470 + 1200$
$= 1,870$ ohms or 1.87 kilohms

It also follows that if three *equal* resistances are connected in series:

$$R_T = 3R$$

Example: What is the total resistance if four 330 ohm resistors are connected in series?

Ans. $R_T = 4 \times 330$
$= 1,320$ ohms $= 1.32$ kilohms

Basic rule to remember: Adding resistors in series *increases* total resistance.

RESISTORS IN PARALLEL

When two or more resistors are connected in *parallel* (Fig. 2-2), the *total* resistance is given by the formula:

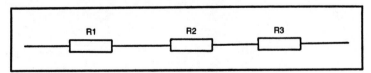

Fig. 2-1. Resistors in series.

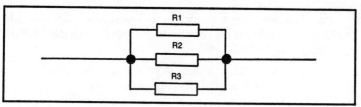

Fig. 2-2. Resistors in parallel.

$$\frac{1}{R_T} = \frac{1}{R_1} + \frac{1}{R_2} \cdots + \frac{1}{R_n}$$

Example: what is the total resistance when a 47 ohm, 15 ohm, and a 22 ohm resistor are connected in parallel?

$$\frac{1}{R_T} = \frac{1}{47} + \frac{1}{15} + \frac{1}{22}$$

$$= \frac{(15 \times 22) + (47 \times 22) + (47 \times 15)}{47 \times 15 \times 22}$$

$$= \frac{330 + 1{,}034 + 705}{15{,}510}$$

$$= \frac{2{,}069}{15{,}510}$$

$$\text{Thus } R_T = \frac{15{,}510}{2{,}069}$$

$$= 7.5 \text{ ohms}$$

This formula becomes cumbersome to work when more than two resistors of large value are involved. However, where *equal* resistor values are involved, calculation is much simplified.

For equal resistance in parallel (Fig. 2-3):

$$\frac{1}{R_T} = \frac{1}{R} + \frac{1}{R} + \frac{1}{R} + \cdots$$

$$= \frac{n}{R}$$

$$\text{or } R_T = \frac{R}{n}$$

16

where n is the number of resistors.

Example: What is the total resistance when four 120 ohm resistors are connected in parallel?

$$\text{Ans. } R_T = \frac{120}{4}$$

$$= 30 \text{ ohms}$$

It follows that a resistance can be halved by connecting a similar value resistor in parallel with it; or reduced to a *third* of its original value by connecting two similar value resistors in parallel with it; and so on. In the case of *two unequal* resistor values connected in parallel, a simplified formula can be used again:

$$R_T = \frac{R1 \times R2}{R1 + R2}$$

or expressed in words $\quad R_T = \dfrac{\text{product of resistor values}}{\text{sum of resistor values}}$

Example: What is the total resistance when a 100 ohm resistor is connected in parallel with a 22 ohm resistor?

$$R_T = \frac{100 \times 22}{100 + 22}$$

$$= 18 \text{ ohms}$$

Basic rule to remember: Adding resistors in parallel *decreases* total resistance. If equal value resistors are used, using two resistors *halves* the total resistance, and so on.

RESISTANCES IN SERIES AND PARALLEL

A circuit may contain resistors in series and parallel. In this case it is first necessary to find the total resistance of each group separately, then treat the two (or more) totals as resistances connected in series.

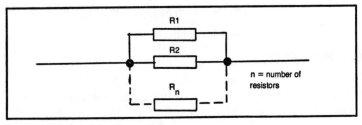

Fig. 2-3. Equal resistors in parallel.

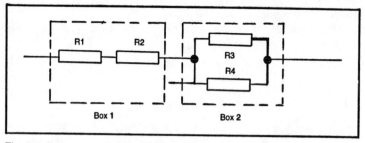

Fig. 2-4. Resistors in series and parallel.

In Fig. 2-4 the two groups are enclosed in 'boxes', the first being resistance in series and the second being resistance in parallel.

$$\text{In the first 'box'} \quad R_{T1} = R1 + R2$$

$$\text{In the second 'box'} \quad R_{T2} = \frac{R3 \times R4}{R3 + R4}$$

The *total* resistance in the circuit is then:

$$R_T = R_{T1} + R_{T2} \text{ (because the 'boxes' are in series)}$$

$$= (R1 + R2) + \frac{R3 \times R4}{R3 + R4}$$

Example: Suppose the resistance values in Fig. 2-4 circuit are R1 = 120, R2 = 270, R3 = 100 and R4 = 68. Find the total resistance.

$$\text{Ans. } R_T = (120 + 270) + \left(\frac{100 \times 68}{100 + 68}\right)$$

$$= 390 + \frac{6,800}{168}$$

$$= 390 + 40.5$$

$$= 430.5 \text{ ohms}$$

DROPPING RESISTORS

A resistor can be used to lower or 'drop' a dc supply voltage to a required volume. In the basic case for calculation normal circuit resistance is represented by a single load resistance (R_L), to which a dropping resistance (R_D) is connected in series (Fig. 2-5).

18

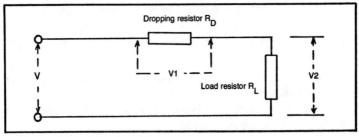

Fig. 2-5. Dropping resistor.

Since the effective resistances are in series the current (I) will be the same throughout the circuit. Thus:

Voltage across R_D or $V1 = I \times RD$)

Voltage across R_L or $V2 = I \times RL$)
$\left(\begin{array}{c}\text{from Ohm's law} \\ E = I \times R\end{array}\right)$

when supply voltage or $V = V1 + V2$.

Example of calculation: Supply voltage is 12 volts dc, and load resistance R_L is 100 ohms. It is desirable to 'drop' the supply voltage so that the voltage across the load is 6 volts dc.

Ans. Since V2 is to be 6, the voltage to be 'dropped' by the dropping resistance is $12 - 6 = 6$ volts, i.e., $V1 = 6$.

Current flow through the circuit is:
$$\frac{V2}{RL}$$
$$= \frac{6}{100}$$
$$= 0.06 \text{ amps}$$

Connecting the circuit as a whole, the *total* resistance in circuit required by Ohm's law is:

$$R_T = \frac{V}{0.06}$$
$$= \frac{12}{0.06}$$
$$= 200 \text{ ohms}$$

Hence the dropping resistance value required is:

$$200 - R_L$$
$$= 200 - 100$$
$$= 100 \text{ ohms}$$

A check calculation will verify if this value is correct:

$$V1 \text{ (or voltage drop across } V_D) = 0.06 \times 100$$
$$= 6 \text{ volts}$$

This verifies that R_D does, in fact, 'drop' the 6 volts—as required.

Another example of the use of a dropping resistor is to convert an ammeter movement into a voltmeter. If V is the voltage range required from the meter and I, is the maximum meter current for full scale deflection of the movement

$$\text{total R required} = \frac{V}{I_i}$$

The value of the series resistance required in this case is $R - R_m$, where R_m is the resistance of the meter movement.

Example: Meter movement has a maximum current of 0.5 amps for full scale deflection and a resistance of 0.1 ohms. Calculate the series resistance required to convert it into a 0 – 10 volt voltmeter.

$$\text{Total resistance required} = \frac{10}{0.5} = 20 \text{ ohms}$$

Therefore the series resistor required = 20 – 0.1 = 19.9 ohms.

Note: the value of meter resistance is commonly negligible compared with the total resistance required in such cases, so series resistance can be calculated directly as $R = \frac{V}{I_i}$

VOLTAGE DROP THROUGH CABLES

Similar principles apply to calculating the voltage drop (voltage loss) through supply cables or wiring. Here in a basic circuit (Fig. 2-6) each cable length is effectively working as a dropping resistor

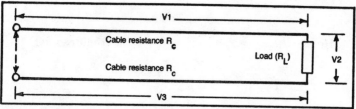

Fig. 2-6. Cable resistance is in series with load.

(although its actual resistance value will normally be quite small). Again the current value is the determining factor in calculation, with effectively three resistors in series, R_c (ingoing), R_L and R_c (outgoing).

Voltage dropped in cable is V1 + V3, where:

$$V1 = IR_c$$
$$V2 = IR_c$$

or total voltage drop in cable $= 2\ IR_c$

The value of I for calculation is determined by the value of the load resistance R_L and the original supply voltage V:

$$I = \frac{V}{R_L}$$

Example: Supply voltage is 22 volts and load resistance is 10 ohms. Calculate the actual voltage available across the load if the resistance of the connecting cable is 0.5 ohms in each load.

Ans. Current through circuit $= \dfrac{22}{10}$

$$= 2.2 \text{ amps}$$

Voltage drop through cables $= I \times R$
$$= 2.2 \times (2 \times 0.5)$$
$$= 2.2 \text{ volts}$$

Therefore actual voltage available across load is 22 − 2.2 = 19.8 volts.

Note: this will imply a slight reduction in current flowing through the load (i.e., total resistance in circuit = 10 + 2 × 0.5 = 11 ohms, giving an actual load circuit of 22/11 = 2 amps).

POTENTIAL DIVIDERS

The basic circuit for a potential divider is shown in Fig. 2-7, where the following voltage apply:

$$V1 = \text{source voltage}$$
$$V2 = \text{output voltage} = \frac{V1}{R1 + R2} \times R1$$
$$V3 = \text{'lost' voltage} = \frac{V1}{R1 + R2} \times R2$$

Note $\dfrac{V1}{R1 + R2}$ is the current flowing through R1 and R2. Also, although V3 is designated 'lost' voltage, it could of course be tapped as a second source of reduced voltage output.

Example: Given a dc source of 20 volts, calculate suitable values of R1 and R2 to give an output voltage (V2) of 15 volts. First it is necessary to nominate a specific value for either R1 or R2 (preferably R1 for simpler calculation), depending on current flow required—i.e., for high currents start with a low resistor value, and vice versa.

For the sake of example, take R1 as 33 ohms. We now have:

$$V2 = \frac{V1}{R1 + R2} \times R1$$

$$\text{or } 15 = \frac{22}{33 + R2} \times 33$$

Solving for R2:

$$33 + R2 = \frac{22 \times 33}{15}$$

$$\text{or } R2 = 48.4 - 33$$

$$= 15 \text{ ohms}$$

This is a practical (available) resistor value, so the calculation is valid. If the calculated value had *not* been a practical value, then the equation would have to be recalculated using different values of R1 as necessary to end up with a practical value for R2.

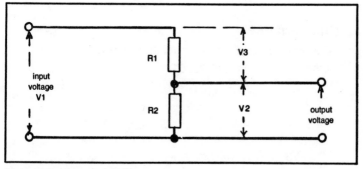

Fig. 2-7. Basic potential divider.

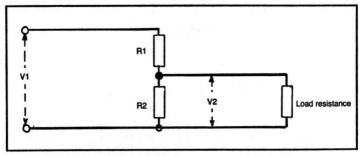

Fig. 2-8. Potential divided with load.

Having arrived at suitable values for R1 and R2, check the current flowing through the circuit.

$$I = \frac{V1}{R1 + R2}$$

$$= \frac{22}{33 + 15}$$

$$= 0.5 \text{ amps (approx.)}$$

If this is too high, (or too low), recalculate the equation starting with a higher (or lower) value of R1.

Note: the actual output voltage and output current will be modified in a complete working circuit by the *load resistance* added to this circuit (Fig. 2-8). Base further calculations for this circuit as done previously.

Chapter 3

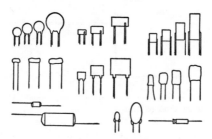

Voltage Drop in Long Leads

Voltage drop produced by the resistance of short leads and inter-connecting wiring is normally negligible. This is not true, however, in the case of long leads—e.g., extension leads. Consult wire data tables (Table 3-1) to determine the resistance of the wire size used for long leads. This is normally expressed in ohms per 1,000 feet of wire. Thus for a twin lead (twin flex):

$$\text{Resistance} = \frac{2 \times L}{1,000} \times R_{1000}$$

where L is the length of lead in feet

R_{1000} is the quoted resistance for that size of wire for 1,000 feet.

The voltage drop follows from the formula:

$$\text{Voltage drop} = \text{current} \times \text{lead resistance}$$

Note: this is valid for both *ac* and *dc* calculations.

Example: 30 gauge wire is chosen for a 100 ft two-wire lead. Resistance for this size of wire is 105 ohms per 1,000 feet (from wire data tables). Supply voltage is 110 V. Load resistance is 100 ohms. Find:

(1) resistance of leads
(2) voltage drop in leads
(3) actual voltage across load

(1) Resistance of leads $= \dfrac{2 \times 100}{1,000} \times 105$

$\qquad\qquad\qquad\quad = 21$ ohms.

Table 3-1. American Wire Sizes (Bare Copper Wire).

Wire size A.W.G. (B. & S.)	Diam. in mils	Circular mil area	Diam. in mm	Feet per pound, bare	Ohms per 1.000 ft 25°C	Current carrying capacity at 700 C.M. per amp	Nearest British S.W.G. no.
1	289.3	83690	7.348	3.947	.1264	119.5	1
2	257.6	66370	6.544	4.977	.1593	94.8	3
3	229.4	52640	5.827	6.276	.2009	75.2	4
4	204.3	41740	5.189	7.914	.2533	59.6	5
5	181.9	33100	4.621	9.980	.3195	47.3	7
6	162.0	26250	4.115	12.58	.4028	37.5	8
7	144.3	20820	3.665	15.87	.5080	29.7	9
8	128.5	16510	3.264	20.01	.6405	23.6	10
9	114.4	13090	2.906	25.23	.8077	18.7	11
10	101.9	10380	2.588	31.82	1.018	14.8	12
11	90.7	8234	2.305	40.12	1.284	11.8	13
12	80.8	6530	2.053	50.59	1.619	9.33	14
13	72.0	5178	1.828	63.80	2.042	7.40	15
14	64.1	4107	1.628	80.44	2.575	5.87	16
15	57.1	3257	1.450	101.4	3.247	4.65	17
16	50.8	2583	1.291	127.9	4.094	3.69	18
17	45.3	2048	1.150	161.3	5.163	2.93	18
18	40.3	1624	1.024	203.4	6.510	2.32	19
19	35.9	1288	.912	256.5	8.210	1.84	20
20	32.0	1022	.812	323.4	10.35	1.46	21
21	28.5	810.1	.723	407.8	13.05	1.16	22
22	25.3	642	.644	514.2	16.46	.918	23
23	22.6	510	.573	648.4	20.76	.728	24
24	20.1	404	.511	817.7	26.17	.577	25
25	17.9	320	.455	1031	33.00	.458	26
26	15.9	254	.405	1300	41.62	.363	27
27	14.2	202	.361	1639	52.48	.288	29
28	12.6	160	.321	2067	66.17	.228	30
29	11.3	127	.286	2607	83.44	.181	31
30	10.0	101	.255	3287	105.2	.144	33
31	8.9	80	.227	4145	132.7	.114	34
32	8.0	63	.202	5227	167.3	.090	36
33	7.1	50	.180	6591	211.0	.072	37
34	6.3	40	.160	8310	266.0	.057	38
35	5.6	32	.143	10480	335	.045	38-39
36	5.0	25	.127	13210	423	.036	39-40
37	4.5	20	.113	16660	533	.028	41
38	4.0	16	.101	21010	673	.022	42
39	3.5	12	.090	26500	848	.018	43
40	3.1	10	.080	33410	1070	.014	44

(2) Total resistance to current is lead resistance plus load resistance = 100 + 21 = 121 ohms.

This current = supply voltage/total resistance

$$= \frac{110}{121}$$

$$= 0.909 \text{ amps}$$

This current is the same through the leads and load resistance, so voltage drop in leads = 0.909 × 21

$$= 19 \text{ volts}$$

(3) Voltage across load thus = 110 − 19 = 91 volts.

WHERE NO WIRE RESISTANCE DATA IS AVAILABLE

If no data is readily available on wire resistances a close approximate value for copper wires can be calculated from the following formula:

$$\text{Resistance per 1,000 ft, } R_{1000} = \frac{.0105}{d^2}$$

25

where d is the bare wire diameter in inches.

Example: Find the resistance of 14 gauge copper wire. Diameter of this size is .010 inches.

$$R_{1000} = \frac{.0105}{(.010)^2}$$
$$= 105 \text{ ohms}$$

This agrees closely with the figure quoted in the previous example.

EFFECT OF TEMPERATURE RISE

The resistance of metallic conductors increases with rise in temperature (e.g., as the wire becomes heated in passing a current). Resistance values for wires are normally quoted for a normal room temperature of 20°C (68°F). Resistance of the same wire at any other temperature is given by:

$$R_T = R_{20} (1 + \alpha (T - 20)$$

when R_{20} is the normal room temperature value (i.e., at 20°C)

α is the *temperature coefficient* of the wire material.

T   is the working temperature

Temperature coefficients are normally quoted for 20°C (68°F). For copper the temperature coefficient is .0393 per °C.

Example: For the same leads, with resistance found to be 21 ohms, find the actual lead resistance if the working temperature is 40°C.

$$\text{Resistance at } 40°C = 21 \times (1 + .0393(40 - 20)$$
$$= 21 \times (1 + .786)$$
$$= 37.5 \text{ ohms}$$

Note: conversion of °F to °C:

$$°F = \frac{9 \times °C}{5} + 32$$

For small ranges in temperature to about 150°F = 66°C a simple approximate formula is:

$$°C = \frac{°F}{2} - 12$$

Example: Convert 100°F into °C:

$$°C = \frac{100}{2} - 12$$
$$= 38$$

This is only an approximate formula but one which can easily be worked out in the head and is sufficiently accurate for most purposes.

Similarly (rough approximate again) °F = 2 × °C + 24

RESISTANCE OF FLEXIBLE LEADS

Resistance and voltage drop of flexible leads can be worked out in exactly the same manner, i.e., using wire table data or calculating resistance from the cross sectional area of copper in the leads. However, the following values can be used for quick and simple calculation.

Flex rating amps	Resistance-ohms per 100 ft	Voltage drop per foot per amp current, mV
3	2.5	25
6	1.7	17
10	1.3	13
15	.95	9.5
20	.55	5.5
25	.335	3.35

Note that resistance here is given per 100 feet of flex (not 1,000 feet), so the resistance formula becomes

$$R = \frac{2 \times L}{100} \times \text{resistance figure above}$$

where L is the actual length of flex used in feet.

The third column in this simple table gives voltage drop direct, knowing the current:

Voltage drop, mV = table figure × L × actual current

Example: Determine the voltage drop in 40 feet of flex with 10 amp rating. This is feeding a 2 kilowatt appliance from a 220 volt supply.

$$\text{watts} = \text{volts} \times \text{amps}$$

$$\text{So current} = \frac{\text{watts}}{\text{volts}}$$

$$= \frac{2,000}{220}$$

$$= 9.09 \text{ amps}$$

$$\text{Voltage drop, mV} = 13 \text{ (from table)} \times 40 \times 9.09$$

$$= 4727 \text{ mV}$$

$$= 4.727 \text{ volts}$$

Chapter 4

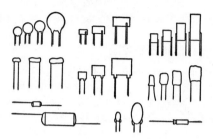

Capacitors (Practical)

Capacitors come in a wide variety of types and physical sizes as well as covering capacity values ranging from 1.8 pF to over 10,000 μF—a numerical range of *millions* of times from the smallest to largest!

Capacitors fall into two broad categories:

☐ Non-polarized types—of which there are many different constructions.

☐ Polarized types which have a definite polarity for working and can be destroyed if connected the wrong way. These are generally known as *electrolytic capacitors* because of their method of construction, (but just to be awkward, there are also non-polarized electrolytic capacitors!).

Non-polarized capacitors consist, basically, of metallic foil interleaved with sheets of solid dielectric material, or equivalent construction. The important thing is that the dielectric is 'ready made' before assembly. As a consequence it does not matter which plate is made positive or negative. The capacitor will work in just the same way, whichever way it is connected in a circuit, hence the description 'non-polarized.' This is obviously convenient, but this form of construction does limit the amount of capacitance which can be accommodated in a single 'package' of reasonable physical size. Up to about 0.1 microfarads, the 'package' can be made quite small, but for capacitance values much above 1 microfarad, the physical size of a non-polarized capacitor tends to become excessively large in comparison with other components likely to be used in the same circuit.

This limitation does not apply in the case of an *electrolytic* capacitor. Here initial construction consists of two electrodes separated by a thin film of *electrolyte*. As a final stage of manufacture a voltage is applied across the electrodes which has the effect of producing a very thin film of non-conducting metallic oxide on the surface of one plate to form the dielectric. The fact that the capacitance of a capacitor increases the thinner the dielectric is made means that very much higher capacities can be produced in smaller physical sizes. The only disadvantage is that an electrolytic capacitor made in this way will have a polarity corresponding to the original polarity with which the dielectric was formed, this correct polarity being marked on the body of the capacitor. If connected the other way in a circuit, the reverse polarity can destroy the dielectric film and permanently ruin the capacitor.

There is also one other characteristic which applies to an electrolytic capacitor. A certain amount of 'unused' electrolyte will remain after its initial 'forming.' This will act as a conductor and can make the capacitor quite 'leaky' as far as *dc* is concerned. This may or may not be acceptable in particular circuits.

CAPACITOR TYPES AND CHARACTERISTICS

Most capacitor types are readily identified by their shape (Fig. 4-1). It is useful to be able to do this because many different types have a similar capacity value range, but some are better than others for specific applications. The following can therefore be regarded as an overall guide to capacitor type selection.

Ceramic Capacitors

These are widely used in miniaturized *af* and *rf* circuits.

They are relatively inexpensive and are available in a wide range of capacities from 1 pF to 1 μF with high working voltages and also characterized by high leakage resistance. They are produced in both disc and tubular shapes; also as metallized ceramic plates.

Disc ceramics. Miniature types may have a maximum voltage rating of only 12 V but can provide large capacity values in a very small package size—e.g., approximately 3 mm diameter for .01μF and 7 mm diameter for .1 μF. Tolerance range can be pretty generous, though—from +50 percent to −20 percent.

General purpose low voltage (50 V rating) disc ceramics are

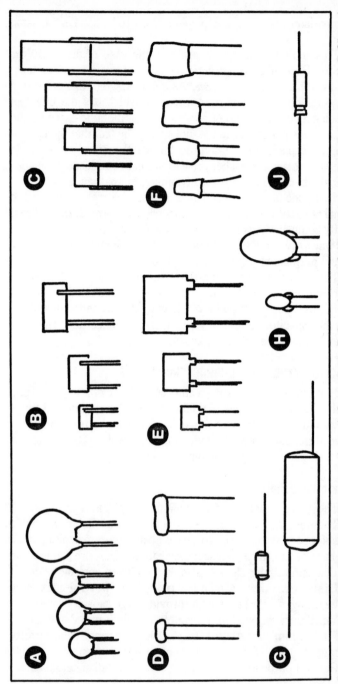

Fig. 4-1. Types of capacitors, approximate actual size; A. Ceramic disc, B. Ceramic plate, C. Monolithic ceramic, D. Tubular ceramic, E. Silvered mica, F. Polyester, G. Polystyrene, H. Tantalum bead, J. Small electrolytic.

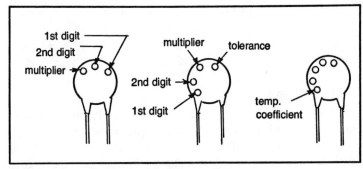

Fig. 4-2. Color coding for ceramic disc capacitors.

slightly larger, with closer positive tolerance possible (+50 percent to −20 percent). Usual value range .01 to .1 μF.

Higher voltage disc ceramics are also available for general purpose use (e.g., with 500 V rating); but in a restricted capacity range—e.g., 10 pF to 10,000 pF. See Fig. 4-2 for color coding.

Tubular ceramics. These have a similar value range to disc ceramics and are again a suitable type for general use. They are obtainable to voltage ratings up to 350 V.

Ceramic plate capacitors. These are (generally) a higher quality type available in a wider range of values in *metallized* ceramic plate (typically from 1.8 pF to 0.1 μF). Physical size ranges from 3.5 mm wide × 4.5 mm high to 6.5 mm wide × 10.5 mm high, size *not* necessarily being related to capacity value. For example, a 1,000 pF plate may be the same size as a 1.8 pF plate; and a 100 pF plate substantially larger.

Metallized plate capacitors are suitable for coupling and decoupling capacitors when stability is not critical.

Monolithic ceramic plate capacitors have much closer tolerances (as good as ± 10 percent) and are particularly suitable for coupling and decoupling applications, and also filters. They can offer high capacitance values (e.g., .001 to .1 μF) in a very small package (e.g., 5 mm square × 2.5 mm thick).

Silvered-Mica Capacitors

These are more expensive than ceramic capacitors but have excellent high frequency response and much smaller tolerances, so are generally regarded as superior for critical *rf* applications. They can be made with very high working voltages. Again these are in 'plate' shape, with capacity values ranging from 2.2 pF up to 10,000 pF. Typical plate sizes are:

2.2 pF to	68 pF	———	13 mm wide ×	8 mm high
82 pF to	220 pF	———	17 mm wide ×	12 mm high
270 pF to	500 pF	———	27 mm wide ×	17 mm high
650 pF to	10000 pF	———	27 mm wide ×	22 mm high

Polystyrene Capacitors

These are made from metallic foil interleaved with polystyrene film, usually with a fused polystyrene enclosure to ensure high insulation resistance. They are noted for their low losses at high frequencies (i.e., low inductance and low series resistance), good stability and reliability. Values range from 10 pF to 100,000 pF, with tolerance as good as ± 1 percent. Case shape is tubular, with a length of approximately 10 mm × 3.5 mm diameter for most values. This type is recommended for tuned circuits, filter networks, discriminators and other control circuits where precision, reliability, stability, and low losses are of major importance.

Polycarbonate Capacitors

These are usually produced in the form of rectangular slabs with wire end connections designed to plug into a printed circuit board. They offer high values of capacity (up to 1μF) in very small sizes, with the characteristics of low losses and low inductance. Like polystyrene capacitors, working voltages become more restricted with increasing capacity value. Polystyrene capacitors are commonly designed specifically for use on printed circuit boards with a typical case size of approximately 7.5 mm × 2.5 mm (larger for values above 0.1 μF); with a pin spacing of 7.5 mm.

Polyester Film Capacitors

These are also designed for use with printed circuit boards, with values from 0.01 μF up to 2.2 μF. Value for value they are generally larger in physical size than polycarbonate capacitors. Their low inherent inductance makes them particularly suitable for coupling and decoupling applications. Values of polyester film capacitors are indicated by a color code consisting of five color bands, reading from the top (see Fig. 4-3). A typical case size is 30 mm wide × 20 mm high.

Mylar Film Capacitors

These can be regarded as a general purpose film type, usually available in values from 0.001 μF up to 0.22 μF with a working voltage up to 100 volts *dc*. Tolerance is usually ± 10 percent.

PETP Film Capacitors

These are a specialized type produced specifically for suppression of electrical interference from domestic appliances. They are generally available in values from .01 μF to .47 μF with working voltages of 250-275 V ac. They are intended to be connected directly across the mains.

Polyester/Paper Capacitors

These are another type of mains capacitor for general use. They are cylindrical in form with the dielectric (polyester film and paper molded in polypropylene. Sizes (and values) range from .001 μF (1 inch long × 9.5 mm diameter to 1 μF (2 inches long × 22 mm diameter).

Electrolytic Capacitors

The original material used for electrolytic capacitors was aluminum foil, together with a paste electrolyte, wound into a tubular form with an aluminum outer cover, characterized by 'dimpled' rings at one or both ends. The modern form of aluminum electrolytic capacitor is based on etched foil construction, enabling higher capacitance values to be achieved in smaller can sizes. Values available range from .1 μF to 47 μF in subminiature sizes and from 1 μF up to 4700 μF (or even larger, if required). Working voltages are generally low, but many range from 10 Vdc up to 250 or 500 Vdc, depending on value and construction. A single lead emerges from each end, but single-ended types are also available (both leads emerging from one end); and can-types with rigid leads in one end for plugging into a socket. Single-ended types are preferred for mounting on printed circuit boards.

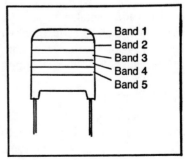

Fig. 4-3. Color coding for polyester film capacitors.

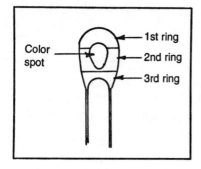

Fig. 4-4. Color coding for tantalum bead capacitors.

Sub-miniature single-ended electrolytic capacitors have a case size of approximately 7 mm long × 4-6 mm diameter. Standard single-ended electrolytics for printed circuit board mainly have case sizes varying from 11 mm long ×5 mm diameter to 33 mm long × 15 mm diameter. Miniature axial lead electrolytics have case sizes ranging from 18 mm long × 8 mm diameter (1 μF) up to 50 mm long × 25 mm diameter (4,700 μF). Can type electrolytics are larger and many range in size up to 3 inches long by 2 inches diameter. These are high value capacitors, the usual range available being from 1,000 μF to 10,000 μF. All electrolytic capacitors have a wide tolerance and choice of value is seldom critical.

Tantalum Capacitors

This is produced both in cylindrical configuration with axial leads, or in *tantalum bead* configuration. Both (and the latter type particularly) can offer very high capacitance values in small physical sizes, within the range 0.1 to 100 μF. Voltage ratings are generally low, e.g., from 35 V down to less than 10 Vdc. Range voltage must not exceed 0.5 V. Typical size of a tantalum bead is 9 × 5 mm for values .1 − 1 μF, up to 17.5 × 9 mm for a 100 μF capacitor.

All electrolytic capacitors normally have their value marked on the body or case, together with a polarity marking (+ indicating the positive lead). Tantalum bead capacitors, however, are sometimes color coded instead of marked with values. This color coding is shown in Fig. 4-4, while other codes which may be found on other types of non-polarized capacitors are given in Figs. 4-5 and 4-6.

Variable Capacitors

Variable capacitors are used for *tuning* or *trimming* to enable the capacitance in a critical circuit to be adjusted. There are physical differences between a tuning capacitor and a trimming capacitor,

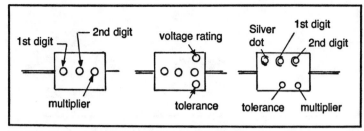

Fig. 4-5. Color coding for molded paper capacitors.

both in size and construction, but both work on the same principle and could, in fact, be used for either duty.

Tuning capacitors are commonly of the air-dielectric type, consisting of two sets of intermeshing metal vanes, one set fixed and the other rotatable by about 180° by means of a spindle (Fig. 4-7). When fully closed (i.e., the spindle is turned so that the rotating vanes are fully in mesh with the fixed vanes) capacitance is at a minimum, and when fully opened (i.e., turned to full movement the opposite way) a maximum.

The maximum capacity of air-dielectric variable capacitors of this configuration ranges from about 10 pF to 1,000 pF. Physical size may range from 'miniature,' e.g., about 25 mm square and 13 mm thick, up to 50 × 60 × 80 mm, or even larger, but is no indication of the working capacity range. It is largely a matter of convenience and compactness to use miniature tuning capacitors for transistor circuits and these may be of the solid-dielectric rather than air-dielectric type. Variable capacitors of this type may be single or *ganged*. The latter are separate capacitors mounted on a common spindle to tune simultaneously.

Trimming capacitors are usually based on a mica-dielectric and consist of two or more plates separated by very thin sheets of mica (Fig. 4-8). Capacity is reduced by turning a central screw to adjust

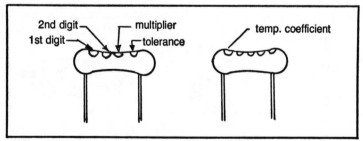

Fig. 4-6. Color coding for phenolic and ceramic tubular capacitors.

Fig. 4-7. Examples of variable capacitors (tuning capacitors).

the pressure between plates and mica. They are designed primarily to be adjusted to a required capacity value and then left set in this position but can, however, be used as a tuning capacitor where their small physical size is an advantage (e.g., in a subminiature receiver). They are also cheaper than tuning capacitors.

Not all such types have a mica-dielectric, some designs being based on a plastic dielectric or even air-spaced. The latter usually provide better response if a trimming capacitor is used as a tuning control.

CAPACITOR CHARACTERISTICS

Like potentiometers, variable capacitors can have different working characteristics, i.e., linear, logarithmic etc., depending primarily on the shape of the plates. These characteristics are usually:

□ *Linear capacitance*, where each degree of spindle rotation produces an equal change in capacitance.

□ *Even frequency,* where each degree of spindle rotation produces an equal change in *frequency* in a tuned circuit.

□ *Square law,* where the change in capacitance is proportional to the *square* of the angle of movement.

□ *Logarithmic,* where each degree of spindle movement produces a constant *percentage* change in frequency.

Fig. 4-8. Trimming capacitors (trimmers).

Table 4-1. Capacitor Parameters.

Capacitor type	Temperature coeff.	Insulation resistance	Power factor (at 1μHz)
Ceramic disc	± 15% /deg. C for monolithic type	more than 10^{10} M ohm	less than 25 × 10^{-2}
Ceramic plate	−150 to −750 ppm /deg C (generally increasing with value)	more than 1,000 M ohm	15 − 350 × 10 (dependent on capacity and size)
Silvered Mica	35 − 75 ppm /deg C	50,000 M ohm	less than 25 × 10^{-4}
Polystyrene	± 160 ppm /deg C	more than 10^{11} M ohm	less than 5 × 10^{-4}
Polycarbonate	−65 ppm /deg C	more than 10^{10} M ohm	less than 30 × 10^{-4}
Polyester	+ 350 ppm /deg C	more than 10^{10} M ohm	less than 150 × 10^{-4}

Linear-frequency characteristics are usually preferred for a radio receiver tuning capacitor.

Like resistors, capacitors also have a temperature coefficient, i.e., change in value with temperature, which may be negative or positive, or 'swing' from negative to positive. *Insulation resistance* and *power factors* can also be significant parameters. Some typical parameter values are shown in Table 4-1.

CHECK LIST OF POSSIBLE TYPES

Table 4-2 indicates the range of capacity values covered by various types of capacitor. Starting with a required capacitor value, check the possible types available. Then refer to previous descriptions to decide on the best types to use.

CAPACITOR VALUES

Working capacitor values range from 1.8 million millionths of a farad (1.8 picofarads or 1.8 pF) up to millionths of a farad (microfarads or μF). Between picofarads and microfarads there are nanofarads (nF) or one thousand millionths of a farad. Confusion often arises because capacitor values are commonly quoted in values from one range overlapping the next. For example:

$$10,000 \text{ pF} = 10 \text{ nF} = .01 \ \mu\text{F}$$

You might find any of these three values given a different circuit diagnosis, all meaning the same value.

The following can be a useful reference to establish such

'overlap,' and also to render the quoted value in the form you are most familiar (or happier working) with. In the above equivalents, for example, most people would prefer to work in .01 μF. At the same time the following list gives values in *preferred numbers*, practical capacitor values following the same rule as resistors in this respect.

pF	nF	μF
(10^{-12} farads)	(10^{-9} farads)	(10^{-6} farads)
1.8		
2.2		
2.7		
3.3		
3.9		
4.7		
5.00		
5.6		
6.8		

Table 4-2. Capacitor Ranges.

pF	nF	μF	Ceramic disc - miniature	Ceramic disc - low voltage	Ceramic disc - high voltage	Ceramic plate (metallized)	Ceramic plate (monolithic)	Silvered-mica	Polystyrene	Polycarbonate	Polyester	Mylar film	Electrolytic - subminiature	Electrolytic - miniature	Electrolytic - can	Tantalum bead	Polyester/paper
1.8								0	0								
to								0	0								
10	1				0	0		0	0	0		0					
	to				0	0	0	0	0	0		0					
10,000	10	.01	0	0	0		0	0	0	0	0	0					
	to				0	0		0	0	0	0						
	100	.1	0			0	0		0	0	0	0	0		0		
		.12						0	0	0	0	0		0			0
		to 1						0	0		0	0		0	0		0
		1.5								0		0	0		0		0
		to										0	0		0		0
		10										0	0		0		0
		15										0	0		0		0
		to										0	0		0		0
		100										0	0		0	0	
		above 100											0				
													0				
		above 1,000													0	0	0
															0	0	0
		above 5,000														0	0	0

38

pF (10^{-12} farads)	nF (10^{-9} farads)	μF (10^{-6} farads)
56		
68		
82		
100	0.1	
120	0.12	
150	0.15	
150	0.18	
220	0.22	
270	0.27	
330	0.33	
390	0.39	
470	0.47	
560	0.56	
680	0.68	
750	0.75	
820	0.82	
1,000	1	.001
1,200	1.2	.0012
1,500	1.5	.0015
1,800	1.8	.0008
2,200	2.2	.0022
2,700	2.7	.0027
3,300	3.3	.0033
3,600	3.6	.0036
3,900	3.9	.0039
4,700	4.7	.0047
5,600	5.6	.0056
6,800	6.8	.0068
8,200	8.2	.0087
10,000	10	.01
12,000	12	.012
15,000	15	.015
18,000	18	.018
22,000	22	.022
27,000	27	.027
33,000	33	.033
39,000	39	.039
47,000	47	.047
56,000	56	.056

pF (10^{-12} farads)	nF (10^{-9} farads)	μF (10^{-6} farads)
68,000	68	.068
82,000	82	.082
100,000	100	.1
	120	.12
	150	.15
	180	.18
	220	.22
	270	.27
	330	.33
	390	.39
	470	.47
	560	.56
	680	.68
	820	.82
	1,000	1
		1.5
		2.2
		3.3
		4.7
		6.8
		10

higher values invariably quoted in μF

Chapter 5

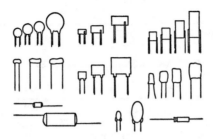

Capacitors (Theoretical)

As far as *dc* is concerned, the capacitor acts as a *blocking* device for current flow (although there will be a certain transient charging current which stops as soon as the capacitor is fully charged). In the case of ac being applied to the capacitor the charge built up during one half cycle becomes reversed on the second half of the cycle, so that effectively the capacitor conducts current through it as if the dielectric did not exist. Thus, as far as *ac* is concerned, a capacitor is a *coupling* device.

There are scarcely any electronic circuits carrying ac which do not incorporate one or more capacitors, either for coupling or shaping the overall *frequency response* of the network. In the latter case a capacitor is associated with a resistor to form an RC *combination* (see later). The charge/discharge phenomenon associated with capacitors may also be used in other types of circuits (e.g., the photographic electronic flash is operated by the charge and subsequent discharge of a capacitor triggered at the appropriate moment).

CAPACITORS IN SERIES AND PARALLEL

For capacitors connected in *series* (Fig. 5-1) the total effective capacity (C) is given by the sum of the reciprocal values of the individual capacitors:

$$C = \frac{1}{C1} + \frac{1}{C2} + \frac{1}{C3} + \dots$$

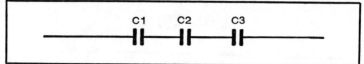

Fig. 5-1. Capacitors in series.

In the case of two *dissimilar* capacitors:

$$C = \frac{\text{product of capacitances}}{\text{sum of capacitances}} = \frac{C1 \times C2}{C1 + C2}$$

In the case of two equal capacitors:

$$C = \frac{C1}{2}$$

Connecting two *similar* capacitors in *series* halves the effective capacity.

For capacitors connected in *parallel* (Fig. 5-2) the total effective capacity is the sum of the individual capacities:

$$C = C1 + C2 + C3 + \ldots$$

In the case of two similar capacitors in parallel:

$$C = 2C1$$

Connecting two *similar* capacitors in parallel doubles the effective capacity.

These rules are the opposite of those for resistors connected in series and parallel.

Capacitance effect is, of course, only apparent in an ac circuit. In a dc circuit a capacitor simply builds up a charge without passing current. In a practical ac circuit a capacitor also exhibits *reactance* and because of its construction may also exhibit a certain amount of *inductance*.

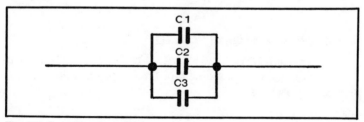

Fig. 5-2. Capacitors in parallel.

REACTANCE OF A CAPACITOR

The reactance of a capacitor, designated X_c (and measured in ohms) is given by the following formula:

$$X_c = 0.16fC$$
where f = frequency in Hz
$$C = \text{capacity in farads}$$
or X_c = .00016fC
where f = frequency in kHz
$$C = \text{capacity in microfarads}$$

BASIC RESISTOR-CAPACITOR (RC) CIRCUITS

A major use of a capacitor is as a coupling device capable of passing *ac* but acting as a block to *dc*. In any practical circuit there will be some resistance connected in series with the capacitor (e.g., the resistive load of the circuit being coupled). This resistance limits the current flow and leads to a certain delay between the application of a voltage to the capacitor and the build-up of charge on the capacitor equivalent to that voltage. It is the 'charge voltage' which blocks the passage of dc. At the same time the combination of resistance with capacitance, generally abbreviated to RC, will act as a *filter* capable of passing ac frequencies, depending on the charge-discharge time of the capacitor, or the *time constant* of the RC combination.

Theoretically the voltage will go on building up indefinitely and never *quite* reach 100 percent of the applied voltage. In practice, the *time constant* is taken as the time for the voltage across the capacitor in an RC circuit to reach 63 percent of the applied voltage, when the following formula applies:

$$\text{time constant T (seconds)} = \frac{R}{C}$$

where R = resistance in ohms
$$C = \text{capacitance in farads}$$
or R = resistance in *megohms*
$$C = \text{capacitance in } microfarads$$

The time constant factor refers to the duration of time in terms of the time factor, e.g., at 1 (which *is* the time factor of the RC combination) 63 percent full voltage has been built up, in a time equal to 2 × the time constant, 80 percent full voltage; and so on. After a time constant of 5 the full (almost 100 percent) voltage will have been built up across the capacitor.

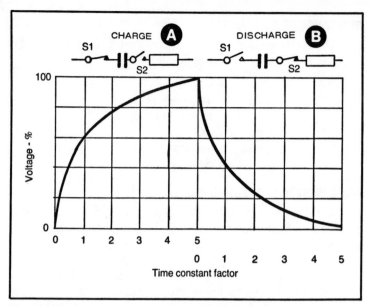

Fig. 5-3. Charge /discharge characteristics of a capacitor.

The discharge characteristics of a capacitor take place in essentially the inverse manner, e.g., after a period of time equal to the time constant the voltage across the capacitor will have dropped to $100 - 63 = 37$ percent of the full voltage; and so on.

In theory a capacitor will never charge up to full applied voltage; nor will it fully discharge. In practice, full charge, or complete discharge, can be considered as being achieved in a period of time equal to five time constants. Thus in the circuit shown in Fig. 5-3A, closing switch 1 will produce a 'full' charge on the capacitor 5x time constant seconds. If switch 1 is now opened, the capacitor will then remain in a condition of storing a voltage equivalent to the original applied voltage, holding this charge indefinitely if there is no internal leakage. In practice it will very slowly lose its charge as no practical capacitor is perfect, but for some considerable time it will remain effectively as a potential source of 'full charge' voltage. If the capacitor is part of a mains circuit, for example, it is readily capable of giving an 'electric shock' at mains voltage if touched for some time after the circuit has been switched off. To complete the cycle of charge-discharge as shown in Fig. 5-3B, Switch 2 is closed. When the capacitor discharges through the associated resistance, it takes a finite amount of time to complete its discharge.

BASIC RC CIRCUITS

A basic RC circuit working with a *dc* input is shown in Fig. 5-4. This circuit will have a time constant dependent on the values of R and C with a corresponding delay in building up the output voltage across C.

Example: Given V = 60 V, R = 5 M ohm and C = .1 μF, find the time constant of this circuit:

$$\text{time constant} = 5 \times 0.1$$
$$= .5 \text{ seconds}$$

This means that after 0.5 seconds (time factor 1), the output voltage will have reached 63 percent of the input voltage, i.e., 60 percent of 60 V or 36 V. After 1 second (time factor 2) it will have reached 80 percent of the input voltage (see Fig. 5-3 again) or 48 volts; and after 1.5 seconds about 90 percent of the input voltage or 54 volts; and so on.

Suppose, for example the output is connected to a neon light which switches on at 50 volts. With this circuit the neon bulb will light up with just a little over 1 seconds delay from the time the input voltage is switched on—and then go out as the capacitor discharges. If the supply voltage is left connected the neon bulb will then flash at this interval or very close to it.

Note that this flashing rate is not the time constant of the RC circuit. It would be, however, if the input voltage was increased so that 63 percent of the input voltage equals the switch-on voltage for the neon bulb (i.e., the neon bulb receives its switch on voltage at a time factor of 1). To do this the input voltage would have to be increased so that:

$$63 \text{ percent } V = 50 \text{ volts}$$
$$\text{or } V = 79.36 \text{ V}$$
$$\text{(approximately 80 volts)}$$

Equally of course, by increasing the input voltage still further

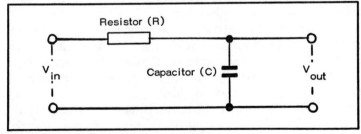

Fig. 5-4. Basic resistor-capacitor (RC) circuit.

the switch on voltage would be reached in a time factor of less than 1. Adjusting the input voltage is thus one way of adjusting the neon bulb's flashing rate in this example. The other way is to play with different values of R and C either by substituting different values calculated to give a different time constant; or with a parallel connected resistor or capacitor. Connecting a similar value resistor in parallel with R, for example, would *halve* the flashing rate (since paralleling similar resistor values halves the total resistance). Connecting a similar value capacitor in parallel with C would *double* the flashing rate. This type of circuit is known as a *relaxation oscillator*. Using a *variable* resistor for R it could be adjusted for a specific flashing rate.

As far as *ac* is concerned, the fact that the applied voltage is alternating means that during one half cycle the capacitor is effectively being charged and discharged with one direction of voltage; and during the second half of the *ac* cycle, charged and discharged with opposite direction of voltage. Thus, in effect, *ac* voltages pass through the capacitor, restricted only by such limitations as may be applied by the RC *time constant* which determines what proportion of the applied voltage is built up and discharged through the capacitor. At the same time a capacitor will offer a certain opposition to the passage of ac through *reactance*, although this does not actually consume power. Its main influence is on the frequency response of RC circuits.

SIMPLE COUPLING

Coupling one stage of a radio receiver to the next stage via a capacitor is common design practice. Although the capacity is apparently used on its own, it is associated with an effective series resistance represented by the 'load' of the stage being fed (Fig. 5-5). This, together with the capacitor, forms an RC combination which will have a particular time constant. It is important that this time constant matches the requirements of the ac signal *frequency* being passed from one stage to the other.

In the case of AM radio stage, the maximum af signal likely to be present is 10 kHz. The 'cycle' time of such a signal is 1/10,000 = 0.1 milliseconds. However, to pass this frequency each cycle represents two charge/discharge functions as far as the coupling capacitor is concerned, one positive and one negative. Thus, the time period for a single charge/discharge function is 0.05 milliseconds. The RC *time constant* necessary to accommodate this needs to be this value to 'pass' 63 percent of the applied *ac*

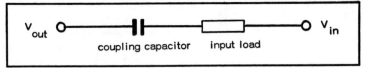

Fig. 5-5. Capacitor as a coupling for *ac*.

voltage—and preferably rather less to 'pass' more than 63 percent of the applied voltage.

These figures can give a clue as to the optimum value of the coupling capacitor to use. For example, the typical input resistance of a low power transistor is of the order of 1,000 ohms. The time constant of a matching RC coupling would be 0.05 milliseconds (see above), giving the requirement:

$$0.05 \times 10 = 1,000 \times C$$
$$\text{or} = 0.05 \times 10^{-9} \text{ farads}$$
$$= 0.50 \text{ pF (or preferably rather less, since this would ensure more than 63 percent voltage 'passed')}$$

In practice, a much higher capacitance value would normally be used, e.g., even as high as 1 μF or more. This will usually give better results, at the expense of efficiency of *ac* (in this case *rf*) transmission. (An apparent contradiction, but it happens to work out that way because the load is reactive rather than purely resistive.) What simple calculation does really show is that capacitance coupling becomes increasingly less efficient with increasing frequency of ac signal when associated with practical values of capacitors used for coupling duties.

BASIC FILTER CIRCUITS

A basic RC combination used as a *filter circuit* is shown in Fig. 5-6. From the input side this represents a resistor in series with a capacitive reactance, with a voltage drop across each component. If the reactance of the capacitor (X_c) is much greater than R, most of the input voltage appears across the capacitor and thus the output voltage approaches the input voltage in value. Reactance is inversely proportional to frequency, however, and so with increasing frequency the reactance of the capacitor decreases, and so will the output voltage (an increasing proportion of the input voltage being dropped by the resistor).

As far as effective passage of ac is concerned there is a critical frequency at which the reactance component becomes so degraded

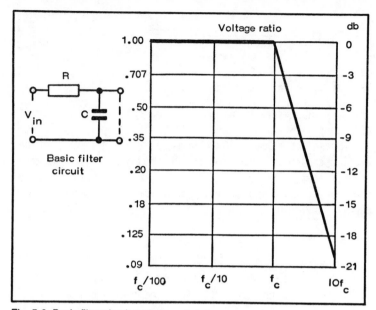

Fig. 5-6. Basic filter circuit and characteristics.

in value that such a circuit starts to become blocking rather than conductive i.e., the ratio of volts$_{out}$ divided by volts$_{in}$ starts to fall rapidly. The critical point, known as the 'roll-off' point or *cut-off frequency* (f_c) is given by

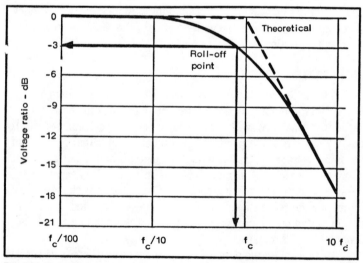

Fig. 5-7. Roll-off point of filters.

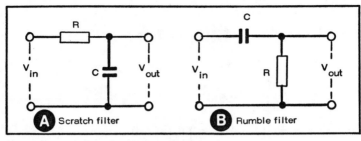

Fig. 5-8. Two basic types of simple filters.

$$f_c = \frac{1}{2\pi\,RC}$$

where R is in ohms

C is in farads

$\pi = 3.1416$

But RC, as noted previously is equal to the time constant of the RC combination:

$$f_c = \frac{1}{2\pi\,T}$$

where T is the time constant, in seconds.

The performance of such a filter circuit is defined by its cut-off frequency and the rate at which the volts$_{in}$/volts$_{out}$ ratio falls above the cut-off frequency. The latter is normally quoted as (so many dB per octave (or each doubling of frequency). Figure 5-7 shows the relationship between dB and volts$_{in}$/volts$_{out}$ ratio; and also the true form of the frequency response curve.

Circuits of this type are called *low-pass filters* because they pass ac signals below the cut-off frequency with little or no loss or attenuation of signal strength. With signals above the cut-off frequency there is increasing attenuation. Suitable component values are readily calculated. For example, a typical *scratch filter* associated with a record player or amplifier used with a record deck would be designed to attenuate frequencies above, say 10 kHz (Fig. 5-8A). This value represents the cut-off frequency required:

$$10,000 = \frac{1}{2\pi\,RC}$$

or RC = 1,600

Any combination of R (in ohms) and C (in farads) giving this product value could be used. A rumble filter, on the other hand, is shown in Fig. 5-8B.

Chapter 6

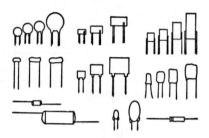

Diodes

A semiconductor diode is a two electrode device which works basically as a rectifier (is conductive connected with one direction of polarity but nonconductive if connected the other way). This is described as forward working (with forward bias) and reverse working (with reverse bias), respectively.

In a practical diode, with forward bias the diode does not start conducting until a specific 'turn on' voltage has been reached. With higher forward voltages the current flow through the diode increases limited only by the maximum voltage of the source and the bulk resistance of the diode. With reverse bias the diode does not 'turn off' completely but has a small leakage current, which is substantially constant regardless of reverse voltage after it has reached its saturation value. Typical diode characteristics are shown in Fig. 6-1.

The construction of a diode governs both its current-carrying capabilities when conducting, and its capacitance effect. The larger the *junction area* of a diode, the higher the current it can pass without overheating—this characteristic being desirable in high power rectifiers, for example. On the other hand, increasing the junction area increases the readiness with which a diode will pass ac due to inherent capacitance effects. To reduce this effect to a minimum, a diode can be made from a single 'doped' crystal (usually N-type), on which the point of a piece of spring wire rests. The end of this wire is given opposite doping (i.e. P-type). This reduces the junction area to a minimum, such a diode being known as a *point-*

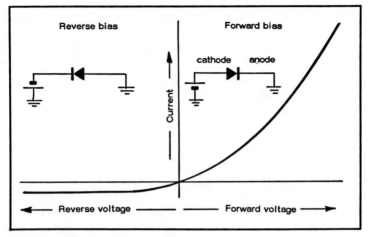

Fig. 6-1. Typical diode characteristics.

contact type. It is a favored type for use in circuits carrying high frequency *ac* signals, and for this reason is sometimes called a *signal diode*.

Diodes may also be described in more general terms by the crystal material (germanium or silicon); and by construction. Here choice can be more important. Germanium diodes start conducting at lower voltages than silicon diodes (about 0.2 to 0.3 volts, as compared with 0.6 volts); but tend to have higher leakage currents when reverse biased, this leakage current increasing fairly substantially with increasing temperature. Thus the germanium diode is inherently less efficient as a rectifier than a silicon diode, especially if reverse bias current is high enough to produce appreciable heating effect. On the other hand, a germanium diode is preferred to a silicon diode where very low 'operating' voltages are involved because it starts to conduct at lower forward voltage.

Maximum permissible junction temperature for diodes is of the order of 90-100°C for a germanium diode and 125-200°C for a silicon diode, depending on construction. The effect of heating (temperature rise) is an 'upward' shift of the current/forward bias curve equivalent to a change in the 'turn on' voltage of about −2mV for degree C.

Typical *reverse leakage currents* for diodes are 2 μA for small germanium diodes and 20 mA for small silicon diodes. With large area power diodes, typical leakage currents are 0.5 mA for germanium diodes and 5 μA for silicon diodes. These figures refer to normal ambient temperature. Reverse current leakage increases

51

with increasing function temperature at an approximate rate of doubling for every 10°C temperature rise.

Diode leakage values are normally quoted for a high temperature (100°C, when leakage values at lower temperatures can be estimated fairly accurately). For example, a leakage figure quoted for 100°C would be halved at 50°C. Unfortunately this simple rule does not work so well the other way. Thus a leakage figure quoted for 50°C would not necessarily be *doubled* at 100°C. It would probably be appreciably less, especially in the case of silicon planar diodes.

DIODES IN AC AND DC CIRCUITS

It will be appreciated that a diode will work in both a *dc* and an *ac* circuit. In a *dc* circuit, it will conduct current if connected with forward bias. If connected the opposite way, it will act as a 'stop' for current flow. An example of this type of use is where a diode is included in a *dc* circuit (say the output side of a *dc* power supply) to eliminate any possibility of reverse polarity voltage surges occurring which could damage transistors in the same circuit.

In an *ac* circuit a diode will 'chop' the applied *ac*, passing half cycles which are positive with respect to the +end of the diode, and stopping those half cycles which are negative with respect to +end of the diode. This is *rectifier* action, widely used in transforming an *ac* supply into a *dc* output. The same action is required of a *detector* in a radio circuit. Here the current applied to the diode is a mixture of *dc* and *ac*. The *diode detector* transforms this mixed input signal into a *varying dc* output, the variations following the form of the *ac* content of the signal.

DIODE CASES

Typical shapes of semiconductor diodes are shown in Fig. 6-2. The positive end (i.e. the end which is to be connected to the positive side of a circuit) is usually marked by a red dot or color band, or a + sign; and also usually with a type number consisting of one or more letters followed by numbers. This identifies the diode

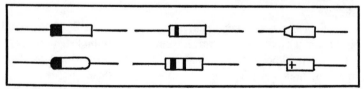

Fig. 6-2. Typical shapes of diodes.

by manufacturer and specific 'model.' Specific type numbers are usually quoted for specific circuit designs, but many circuits are fairly non-critical as regards the type of diode used. In circuit designs, the P side or anode of a diode may be marked A (or a), and the N side or cathode marked K (or k).

DIODE CATEGORIZATION

Diodes may be generally described as germanium diodes, silicon diodes, silicon planar diodes, etc., also as general purpose, af or rf signal diodes, power diodes, silicon rectifier diodes, switching diodes, fast logic or high speed diodes, etc. Further types are described by name, e.g. breakdown diodes (DIACS), Zener diodes, variable capacitance diodes, Schottky diodes, photodiodes, light emitting diodes, etc.

DIODE CHARACTERISTICS

Characteristic values *normally* quoted for diodes are:

☐ *Maximum reverse voltage*—with figures normally given for *peak* (or absolute maximum which the diode will tolerate) and *average*. Separate values may also be given for different temperatures, i.e., ambient (usually 20° centigrade or 25° centigrade). In the absence of separate temperature ratings ambient temperatures can be inferred.

☐ *Maximum forward current*—given in milliamps. Again separate figures may be given for *peak* and *average*; and separate values for ambient temperature and 60° centigrade.

☐ *Forward Voltage* (Vf)—or 'turn on' voltage with a corresponding forward current.

☐ *Maximum leakage current*—with reverse bias, at a specified voltage.

☐ *Ambient temperature range*—referring to the maximum and minimum ambient temperatures, within which range the diode will not be harmed.

☐ *Maximum junction temperature*—usually separate figures for continuous and intermittent operations.

☐ *Thermal resistance*—in °Centigrade per millivolt.

DIODE EQUIVALENTS

Like transistors there are many different types of diodes—but far less numerous in actual numbers. Use of a specific type or type number is not critical in many circuits, but in others it can be. Thus

Table 6-1. Diode Equivalents.

(The diode type listed on the right is a direct replacement for the diode in the left-hand column)

BA100	BAX16	OA85	OA95
BA145	BY206	ZS170	IN4001
BA148	BY206	ZS171	IN4002
BAY38	BAX16	ZS172	IN4003
BY100	IN4006	ZS174	IN4004
DD000	IN4001	ZS175	IN4005
DD001	IN4002	ZS178	IN4006
DD003	IN4003	ZS270	IN5400
DD006	IN4004	ZS271	IN5401
IGP7	OA90	ZS272	IN5402
ISJ50	OA200	ZS274	IN5404
ISJ150	OA202	ZS276	IN5406
OA70	OA90		
OA73	OA90		
OA79	AA119		
OA81	OA91		

for many basic circuits only a germanium (or silicon) diode may be specified, when virtually any general purpose germanium (or silicon) diode would probably do—but not substituting a germanium diode for a silicon diode. With rather more critical circuits, an 'equivalent' diode should be used, if the original diode specified is not readily obtainable. Some useful diode equivalents are given in Table 6-1.

BASIC DIODE CIRCUIT

The basic diode circuit is shown in Fig. 6-3, comprising a diode in series with a load resistance and an input signal source V_{in}. Voltage V_d across the diode is given by:

$$V_d = V_{in} - IR_L$$

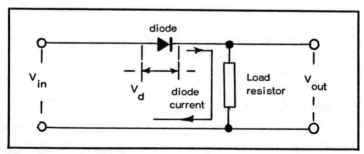

Fig. 6-3. Basic diode circuit.

where I is the current flowing through the circuit and is the source through the diode and the load resistance.

This formula can be used to establish the *load line* on the diode characteristic curve. Taking the two extreme points, viz:

when current $I = 0$, $V = V_{in}$
when $V_{in} = 0$, current $I = V_{in}/R_L$

The working part of the diode for an instantaneous value of V_{in} is then established by the part where the load line cuts the characteristic curve - Fig. 6-4.

In practice there can be a slight snag in attempting to do this. The current value calculated for the position $V_{in} = 0$ can come way above the scale height of the diode characteristic curve, as given by the manufacturer. In this case choose an arbitrary value for current I_a, which does come on the scale. At this point the corresponding arbitrary voltage is:

$$V_a = V_{in} - I_a R_L$$

This establishes the position of the arbitrary point, so enabling the load line to be drawn as a straight line from there to $V = V_{in}$ on the bottom line ($I = 0$).

This establishes the *static* characteristics of the basic diode circuit. If the input (V_{in}) varies, the effective load line will also shift upwards or downwards, but retaining the same shape (since this is governed by R_L). Thus to determine the *dynamic* characteristics it is necessary to recalculate a current value for each voltage value,

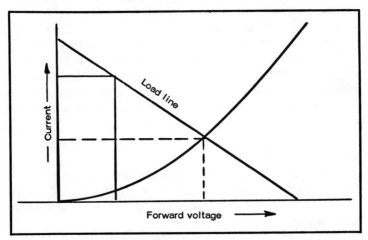

Fig. 6-4. Working point is established where load line cuts characteristic diode curve.

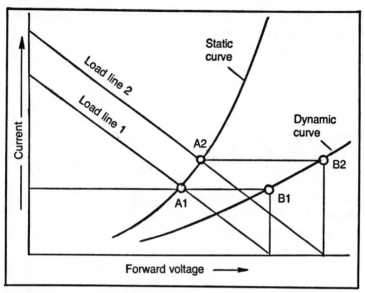

Fig. 6-5. Derivation of the dynamic curve.

which will have the effect to shifting the characteristics to the right Fig. 6-5.

Example: Suppose for simplicity we consider just two values V1 and V2 of a varying input voltage, with a load resistance of R_L. Each will represent a different load line, running from V1 to $I = V1/R_L$, and V2 to $I = V2/R_L$, respectively. Each load line establishes a different working point, A1 and A2 on the *static* curve.

Starting with load line 1, project lines vertically from V1 and horizontally through the static working point. Where these meet B1 establishes a point on the *dynamic* curve. Similar treatment at point V2 on the horizontal scale establishes another point B2 on the *dynamic* curve. Repeat at other values of the varying input, V3, V4, etc, as necessary to be able to join up all 'B' points and complete the *dynamic curve*.

LOAD RESISTOR VALUES

The load resistor R_L value must be chosen to limit the diode forward current to no more than its maximum rated value. The diode itself also contributes some resistance, but this can be ignored for practical purposes.

$$\text{Calculate } R_L \geq \frac{\text{maximum input voltage}}{\text{maximum rated forward current}}$$

Example: The diode chosen is rated for a maximum forward current of 200 mA. Maximum input voltage is 12V.

$$R_L \geq \frac{12}{.200}$$

$$\geq 60 \text{ ohms}$$

The nearest preferred value to use is 68 ohms. 82 ohms or 100 ohms might be better, provided the resulting smaller current is within the required dynamic working range (e.g., using the diode as a clipper circuit).

TRANSFER CHARACTERISTICS

The relationship between the output voltage V_o to the input voltage V_{in} is called the *transfer characteristics* and this can be determined from the dynamic curve, as in Fig. 6-6. The dynamic curve simply has its vertical (current) axis re-designated in terms of corresponding voltage (corresponding currents multiplied by R_L), when it becomes a *transfer curve*.

The input waveform (shown as a triangular wave for simplicity of illustration) is then plotted on a vertical axis below the transfer curve. Taking various points along this curve, projects upwards to meet the dynamic curve, then transfer this value horizontally to form points for plotting the output curve.

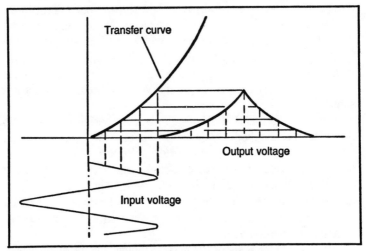

Fig. 6-6. Derivation of output voltage from input voltage and transfer curve (diagrammatic only).

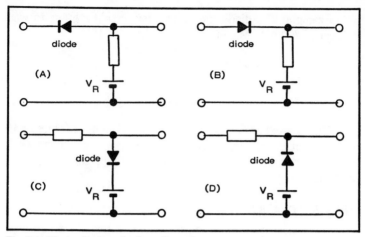

Fig. 6-7. Diode clipping circuit configurations.

Two points to note are:

☐ The diode acts as a 'clipper', i.e., cuts off input to give no output up to a specific positive value of input signal voltage.

☐ Since the transfer curve is not a straight line but a curve, the output is also non-linear, i.e., distorted slightly.

DIODE CLIPPING CIRCUITS

Practical diode clipping circuits are based on providing a positive cut-off or *breakpoint* rather than an arbitrary point determined by the transfer characteristics of the diode. At the same time the whole of the output can then be derived from a linear region of the transfer curve, so eliminating distortion. This can be done by providing a constant regulating voltage V_R in the basic diode in series with a resistor circuit. Four possible arrangements for this are shown in Fig. 6-7.

In Fig. 6-7A and B, the diode appears as a series element, with the polarity of the diode determining whether the top or bottom of the input waveform is clipped. Performance is readily predictable from the transfer characteristics of the diode—i.e., establishing a suitable value for the regulating voltage V_R. The disadvantage of this configuration is that in the 'off' condition the diode may still be capable of transmitting high frequency waveforms via the inherent capacitance of the diode.

In C and D the resistor is a series element and the diode is now a shunt element. This avoids the problem of high frequency transmission through the diode in the 'off' condition, but also has disadvantages. For one, the impedance of the source which supplies V_R must be kept low (a requirement which does not arise when the resistor is a shunt element since the value of this resistor is normally much higher than the source impedance). The second, and more significant, disadvantage is that the diode capacitance in shunt configuration will tend to round sharp edges of input waveforms (e.g. square waves or sawtooth waves) and also attenuate high frequency signals. This effect will be aggravated by any other capacities in shunt with the output terminals.

A diode circuit for clipping at two separate levels is shown in Fig. 6-8, the diodes appearing in shunt configuration. Each diode has its own separate regulating voltage supply, the requirement being that V_{R2} should be greater than V_{R1}. The output signal is then clipped at V_{R1} and V_{R2} as shown.

Working characteristics of this circuit are:

Input	Output V_o	diode state	
		D1	D2
$V_{in} \leq V_{R1}$	V_{R1}	on	off
$V_{R1} < V_{in} < V_{R2}$	V_{in}	off	off
$V_{in} \geq V_{R2}$	V_{R2}	off	on

Again, suitable values for V_{R1} and V_{R2} can be established from the transfer characteristics diagram for any particular input waveform. A particular application of this circuit is to connect a sinusoidal input waveform into a (near) square wave output.

DIODES AS CLAMPS

A basic *clamping circuit* is shown in Fig. 6-9. Here, regardless of the make of the input voltage the output voltage cannot rise above V_R. In other words, the output is *clamped* to V_R.

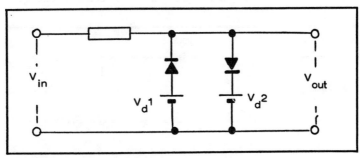

Fig. 6-8. Diode clipping at two different levels.

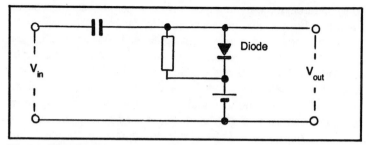

Fig. 6-9. Diode clamp.

With a sinusoidal input voltage having a peak value of V_p and an average of 0, the effect of clamping the output to V_R is to reproduce the input waveform at the output with the same peak value, but now having an average of $V_R - V_p$, provided the time constant RC of the circuit is much larger than the period of the signal. Suitable values for R and C must thus be selected accordingly.

In practice, due to the fact that the diode will have a positive 'turn on' voltage and some finite resistance, and the source will bias some impedance, perfect clamping is not likely to be obtained from such a simple (uncompensated) circuit. In other words, the output voltage may rise slightly above V_R, and there may be some distortion in the output waveform. These effects should be negligible in non-critical circuits.

DIODES AS RECTIFIERS

A basic diode rectifier circuit is shown in Fig. 6-10. The diode conducts only when V_{in} is more positive than V_{out}, so that with a pulse-wave or sine-wave input, waveform conduction is initiated until using V_{in} up to the full V_{in} value. At the same time the capacitor C is charged through R_1 and the diode. It may or may not reach a charge of $+V_{in}$ before V_{in} peaks and starts to descend, depending on

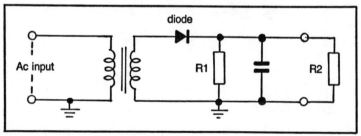

Fig. 6-10. Basic diode rectifier circuit.

the RC time constant. As soon as V_{in} has fallen below V_{out}, capacitor C discharges through R2.

Although a simple circuit, detailed performance analysis is difficult and the output characteristics can vary considerably. If R1 is made a small value then V_{out} is a *dc* voltage approximately equal to the positive input peak voltage for any waveform, but with marked variation or 'ripple' because of resistive losses present in a practical circuit. The theoretical value of the output voltage is:

$$\frac{V_{in} \times R1}{R1 + R2}$$

If R1 is made so large that CR is substantially greater than the cycle time of the input voltage, V_{out} is equal to the positive average value of V_{in} (i.e., a somewhat lower nominal value, but smoother). Specifically, this type of circuit is a *half-wave* rectifier since it transforms only the positive halves of the input voltage into dc output, ignoring (blocking) the negative halves.

The full wave rectifier circuit of Fig. 6-11 overcomes this limitation, at the expense of requiring a transformer to produce a push-pull input. The value of C only needs to be one half that used in a half-wave rectifier circuit, since it discharges through R for only half the signal period before being recharged. Another advantage is that the ripple frequency is twice the input frequency, making it easier to remove. This depends strictly on the values of R1 and R2 being equal, and the characteristics of the two diodes being identical. If not a ripple component will be introduced at the input frequency as well as at twice that frequency.

The preferred full-wave rectifier circuit is the bridge rectifier, shown in Fig. 6-12. This dispenses with the need for transformer coupling and duplication of resistors on the input side, and the degree of ripple is almost entirely related to differences in the

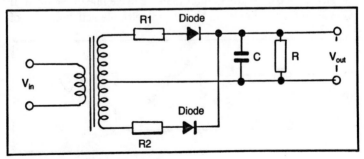

Fig. 6-11. Full wave rectifier circuit.

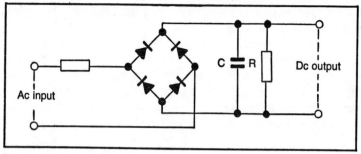

Fig. 6-12. Bridge rectifier circuit.

characteristics of diodes used. These should be matched as closely as possible.

This bridge circuit can also be used with transformer coupling, of course, but the transformer can be smaller (and cheaper), and does not require a center tap. Normally transformer coupling would only be used to step-up or step-down the initial *ac* voltage to an acceptable level, or to provide *dc* isolation between input and output.

DC RESTORER

With the components of the basic half wave rectifier circuit rearranged so that the capacitor C and diode are interchanged (Fig. 6-13), the circuit works as a *DC restorer*. The output consists of a *dc* component equal to peak V_{in} (C very large) or average value of V_{in} (C small); together with an *ac* component identical to V_{in}. The combined result is an output waveform just reaching zero potential at its positive peaks (or just reaching zero potential at its negative peaks if the diode is connected the other way).

Practical design features of this circuit are that R1 should have as low a value as possible, with R2 very much greater in value. CR1 must also be substantially greater than the cycle line of the input. If C is made too low, there will be noticeable ripple and distortion of

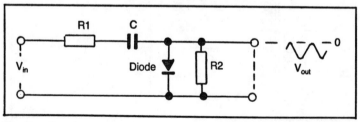

Fig. 6-13. Dc restorer.

the output waveform. If C is made too high, many cycles of input can pass before the restoration level at the output is reached.

ZENER DIODES

A *Zener diode* is a special type of silicon junction diode with a low and *constant* breakdown voltage. Typical working characteristics are shown in Fig. 6-14. The sharp break from non-conducting to almost perfect conductance at a particular reverse voltage is called the zener knee. This characteristic makes a zener diode a particularly useful device for producing voltage-stabilizing or current-stabilizing circuits. A zener diode is normally operated at or above breakdown voltage, with sufficient resistance in the circuit to limit the actual current flowing in the circuit to a safe figure for the zener diode used.

In the basic circuit shown in Fig. 6-15, a zener diode is connected in a series with a limiting resistor to a supply voltage. This resistor effectively divides the supply voltage into the breakdown voltage across the diode with the remainder dropped across the resistor. The voltage across the zener diode *remains* constant, even if the supply voltage varies. Thus tapping the circuit at this point can provide a constant voltage output. The only thing that happens with a variable supply voltage is that the current flow will rise or fall in proportion. Only if the supply voltage falls so much that the working

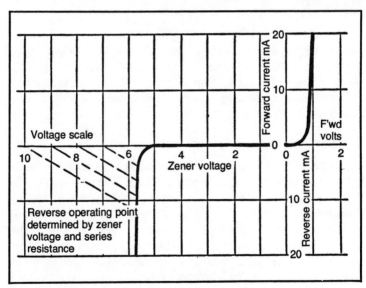

Fig. 6-14. Working characteristics of a typical zener diode.

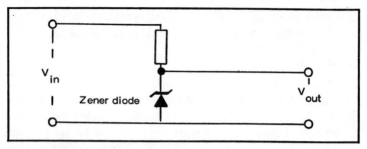

Fig. 6-15. Basic circuit for a zener diode.

point of a diode is pulled back past its knee point will the constant voltage output cease.

Only one constant voltage can be tapped from the zener diodes, corresponding to the *breakdown voltage* or *zener voltage* as it is generally called. Typically this is of the order of 5 to 6 volts. Being a small device it is also obvious that the power it can provide in a constant voltage circuit is also limited, usually to the order of milliwatts only. Zener diodes are, however, produced with much larger zener voltages and also power capacities of the order of watts.

Typical outlines (can shapes) for zener diodes are:

☐ D07   power ratings up to 400-500 mW
☐ DO29   1 watt
☐ DO1 and DO3   1.5 watt
☐ DO4   20 watt

Polarity is defined by the cathode identified by a colored band or chamfered end; or in the case of higher power zeners (DO1, DO3, and DO4 casings) by connection of the cathode to the envelope or a stud. *Connection note:* for zener operation the cathode is made positive.

BASIC ZENER DIODE STABILIZER

A primary application of a zener diode is to obtain a stable voltage source from a supply which may be subject to variations. In this case the required (stabilized) voltage is taken from across the zener diode (as shown in Fig. 6-15). This derived supply voltage will be the zener voltage for that particular diode, as quoted in the specification. However, since we are talking of a *stabilized* supply, due consideration needs to be given to *unstabilizing* factors which may be present due to variations of the zener diode characteristics in a working circuit. Specifically these are:

□ Tolerance on the specified zener voltage

□ Zener temperature coefficient (affecting the resistance and zener current)

□ Tolerance of R in the circuit (again affecting zener current)

□ Changes in input voltage (again affecting zener current)

Whether these need to be analyzed in detail or not depends on how critical the source voltage requirement is. For most purposes, the last three points above can be ignored since they will, to some extent, be offsetting (i.e., tend to cancel each other out). By far the biggest variation is likely to be the initial tolerance of the zener. However, this will not change, i.e., the zener will have a V_2 value within the specified tolerance, be it 10 percent above or below the specified V_2 at likely extremes. This will merely establish the source voltage at a specific value above or below the specification figure.

If source requirements are critical, then rather than work out possible variations in detail it is easier to adopt a more refined stabilizer circuit, such as that shown in Fig. 6-16. Here the earthy side of the load is returned to a potential divider R2 R3. Now if V_{in} rises causing the positive output terminal to rise, the negative terminal also rises. Provided the rise is equal on each terminal, perfect stabilization is maintained. This can be achieved if:

$$\frac{R3}{R2} = \frac{Rz}{R1}$$

In practice stabilization provided by such a circuit can never be perfect because R1, R2 and R3 will not have exact values, and Rz is a function of the zener current. Nevertheless it can be a very effective circuit if the load resistance is constant and the variations in the input voltage V_{in} are not more than about ± 10 percent.

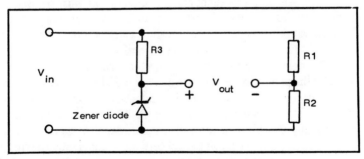

Fig. 6-16. Stabilized voltage source using zener diode.

Zener diodes can also be used in clipper circuits, and as coupling devices. A basic rule to observe in such cases is that the zener diode should not be operated at currents below the standard quoted level, which is commonly 5 mA. If they are they can prove very 'noisy' particularly if a capacitor is connected in parallel with the zener.

VARIABLE CAPACITY DIODES

Variable capacity diodes, known as *varicaps* or *varactors*, are a special type of diode which behave as a capacitor at high Q when biased in the reverse direction, the actual capacitance value being dependent on the bias voltage applied. Typical applications are for the automatic control of tuned circuits or 'electronic tuning', adjusting capacity in the circuit, and thus resonant frequency, in response to changes in signal voltage; automatic frequency control of local oscillator circuits in superhets and TV circuits; and also as frequency doublers and multipliers.

TUNNEL DIODES

The *tunnel diode* is another type with special characteristics, unlike that of any other semiconductor device. It is constructed like an ordinary diode but the crystal is more heavily doped, resulting in an extremely thin barrier (potential layer). As a consequence, electrons can *tunnel* through this barrier.

This makes the tunnel diode a good conductor with both forward *and* reverse voltage. Behavior, however, is quite extraordinary when the forward voltage is increased (see Fig. 6-17). Forward current at first rises with increasing forward voltage until it reaches a peak value. With increasing forward voltage it then drops, to reach a minimum or *valley* value, I_v. After that it rises again with further increase in forward voltage. Worked in the region from peak voltage to valley voltage, the tunnel diode exhibits *negative resistance* characteristics. Another interesting feature is that any forward current value between peak and valley value is obtainable three times (at different forward voltages).

The negative resistance ($-R_n$) of a tunnel diode has a minimum value at the point of inflexion between I_p and I_v. Its series resistance R_s is pure ohmic resistance. Series inductance L_s depends on the design of the package and the lead length. The function capacitance C, normally measured at the valley point, depends on the bias.

Typical values of these parameters for a tunnel diode having a peak current value I_p of 10 mA are:

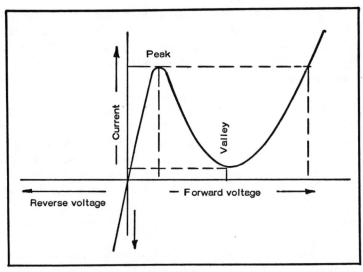

Fig. 6-17. Characteristics of a tunnel diode.

$$-R_n = 30 \text{ ohms}$$
$$R_s = 1 \text{ ohm}$$
$$L_s = 5 \text{ nH}$$
$$C = 20 \text{ pF}$$

Tunnel diodes have a useful application for very high speed switching, with a particular application to pulse and digital circuitry, e.g., digital computers.

SCHOTTKY DIODE

This is a metal semiconductor diode, formed by integrated circuit techniques and generally incorporated in ICs as a 'clamp' between base and emitter of a transistor to prevent saturation. Voltage drop across such a diode is less than that of a conventional semiconductor diode for the same forward current. Otherwise its characteristics are similar to that of a germanium diode.

PHOTODIODES

It is a general characteristic of semiconductor diodes that if they are reverse biased and the junction is illuminated, the reverse current flow will vary in proportion to the amount of light. This effect is utilized in the *photodiode* which has a clear window through which light can fall on one side of the crystal and across the junction of the P- and N-zones.

In effect, such a diode will work in a circuit as a *variable resistance*, the amount of resistance offered by the diode being dependent on the amount of light falling on the diode. In the dark the photodiode will have normal reverse working characteristics, i.e., provide almost infinitely high resistance with no current flow. At increasing levels of illumination, resistance will become proportionately reduced, thus allowing increasing current to flow through the diode. The actual amount of current is proportionate to the illumination only, provided there is sufficient reverse voltage. In other words, once past the 'knee' of the curve, the diode current at any level of illumination will not increase substantially with increasing reverse voltage (Fig. 6-18). It will be noticed that there is a separate curve for different levels of illumination. None of these curves pass through the original (except the dark current curve corresponding to zero level of illumination).

Photodiodes are extremely useful for working as light operated switches. They have a fairly high switching speed, so they can also be used as counters, e.g., counting each interruption of a beam of light as a pulse of current.

There are two other types of light-sensitive diodes—the photovoltaic diode and the light emitting diode (LED). The *photovoltaic diode* generates an emf or voltage when illuminated by light,

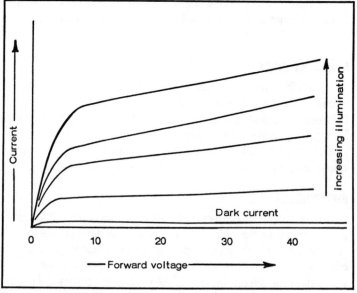

Fig. 6-18. Characteristics of a photodiode.

the resulting current produced in an associated circuit being proportional to the intensity of the light. This property is utilized in the construction of light meters. The amount of current produced by a photodiode can be very small, and so some amplification of the current may be introduced in such a circuit. Special types of photodiodes, constructed more like diode valves, are known as photocells and are generally more suitable for use as practical light meters. The *light emitting diode* works the opposite of a photovoltaic diode. It emits light when a current is passed through it.

Chapter 7

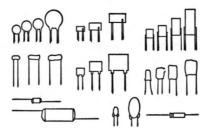

Transistors (Introduction)

Put very simply a semiconductor material is one which can be 'doped' to produce a predominance of electrons or mobile negative charges (n-type); or 'holes' or positive charges (p-type). A single crystal of germanium or silicon treated with both n-type dope and p-type dope forms a semiconductor *diode*, with the working characteristics described in Chapter 6. Transistors are formed in a similar way but like two diodes back-to-back with a common middle layer doped in the opposite way to the two end layers, thus the middle layer is much thinner than the two end layers or *zones*.

Two configurations are obviously possible, pnp or npn (Fig. 7-1). These descriptions are used to describe the two basic types of transistors. Because a transistor contains elements with two different polarities (i.e., 'p' and 'n' zones), it is referred to as a bipolar device, or *bipolar transistor*.

A transistor thus has three elements with three leads connecting to these elements. To operate in a working circuit it is connected with two external voltage or polarities. One external voltage is working effectively *as* a diode. A transistor will, in fact, work as a diode by using just this connection and forgetting about the top half. An example is the substitution of a transistor for a diode as the detector in a simple radio. It will work just as well as a diode as it *is* working as a diode in this case.

The diode circuit can be given forward or reverse bias. Connected with forward bias, as in Fig. 7-2, drawn for a pnp transistor, current will flow from p to the bottom n. If a second voltage is

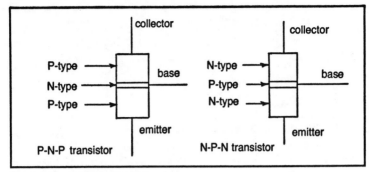

Fig. 7-1. Pnp and npn transistor doping.

applied to the top and bottom sections of the transistor, with the *same* polarity applied to the bottom, the electrons already flowing through the bottom n section will promote a flow of current through the transistor bottom-to-top.

By controlling the degree of doping in the different layers of the transistor during manufacture, this ability to conduct current through the second circuit through a resistor can be very marked. Effectively, when the bottom half is forward biased, the bottom section acts as a generous source of free electrons (and because it emits electrons it is called the *emitter*). These are collected readily by the top half, which is consequently called the *collector*, but the actual amount of current which flows through this particular circuit is controlled by the bias applied at the center layer, which is called the *base*.

Effectively, therefore, there are two separate 'working' circuits when a transistor is working with correctly connected polarities (Fig. 7-3). One is the loop formed by the bias voltage supply encompassing the emitter and base. This is called the *base* circuit or *input* circuit. The second is the circuit formed by the collector voltage supply and all three elements of the transistor.

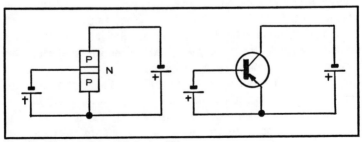

Fig. 7-2. Pnp transistor with forward bias.

71

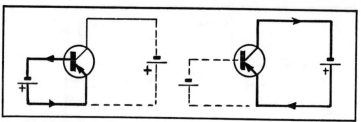

Fig. 7-3. 'Working' circuits in a pnp transistor.

This is called the *collector* circuit or *output* circuit. (Note: this description applies only when the emitter connection is common to both circuits—known as *common emitter* configuration.) This is the most widely used way of connecting transistors, but there are, of course, two other alternative configurations—*common base* and *common emitter*. But, the same principles apply in the working of the transistor in each case.

The particular advantage offered by this circuit is that a relatively small *base* current can control and instigate a very much larger *collector* current (or, more correctly, a small input *power* is capable of producing a much larger *output* power). In other words, the transistor works as an *amplifier*.

With this mode of working the base-emitter circuit is the input side; and the emitter through base to collector circuit the output side. Although these have a common path through base and emitter, the two circuits are effectively separated by the fact that as far as polarity of the base circuit is concerned, the base and upper half of the transistor are connected as a *reverse biased* diode. Hence there is no current flow from the base circuit into the collector circuit.

For the circuit to work, of course, polarities of both the base and collector circuits have to be correct (forward bias applied to the base circuit, and the collector supply connected so that the polarity of the common element (the emitter) is the same from both voltage sources). This also means that the polarity of the voltages must be correct for the type of transistor. In the case of a pnp transistor as described, the emitter voltage must be *positive*. It follows that both the base and collector are negatively connected with respect to the emitter. The symbol for a pnp transistor has an arrow on the emitter indicating the direction of *current flow*, always *towards* the base. ('p' for positive, with a pnp transistor).

In the case of an npn transistor, exactly the same working principles apply but the *polarities* of both supplies are reversed (Fig. 7-4). That is to say, the emitter is always made negative relative to base and collector ('n' for negative' in the case of an npn transistor). This is also inferred by the reverse direction of the arrow on the emitter in the symbol for an npn transistor, i.e., current flow *away* from the base.

While transistors are made in thousands of different types, the number of *shapes* in which they are produced is more limited and more or less standardized in a simple code—TO (Transistor Outline) followed by a number.

TO1 is the original transistor shape—a cylindrical 'can' with the three leads emerging in triangular pattern from the bottom. Looking at the base, the upper lead in the 'triangle' is the *base*, the one to the right (marked by a color spot) the *collector* and the one to the left the *emitter*. The collector lead may also be more widely spaced from the base lead than the emitter lead.

In other TO shapes the three leads may emerge in similar triangular pattern (but not necessarily with the same positions for base, collector and emitter), or in-line. Just to confuse the issue there are also sub-types of the same TO number shape with different lead designations. The TO92, for example, has three leads emerging in line parallel to a flat side on an otherwise circular 'can' reading 1,2,3 from top to bottom with the flat side to the right looking at the base.

With TO92 sub type a (TO92a):   1 = emitter
                                                           2 = collector
                                                           3 = base
With TO92 sub-type b (TO92b):   1 = emitter
                                                           2 = base
                                                           3 = collector

To complicate things further, some transistors may have only two emerging leads (the third being connected to the case internally); and some transistor outline shapes are found with more than three leads emerging from the base. These, in fact, are integrated

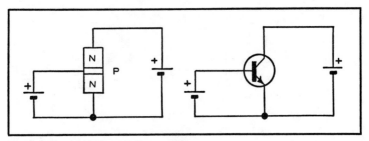

Fig. 7-4. Npn transistor with forward bias.

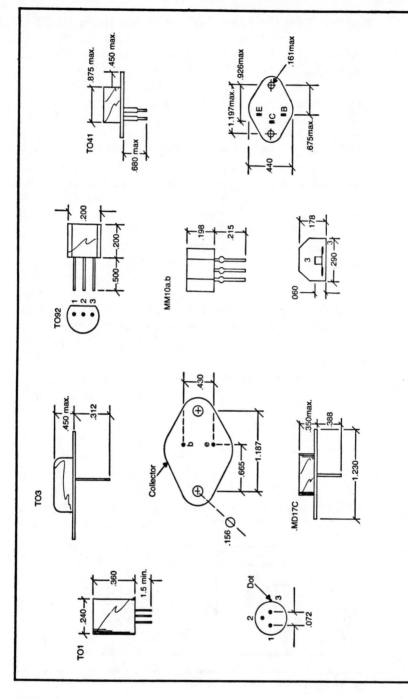

74

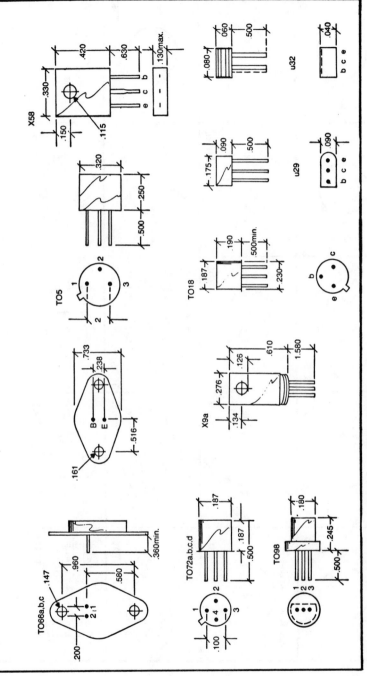

Fig. 7-5. Typical transistor can shapes (dimensions in millimeters).

75

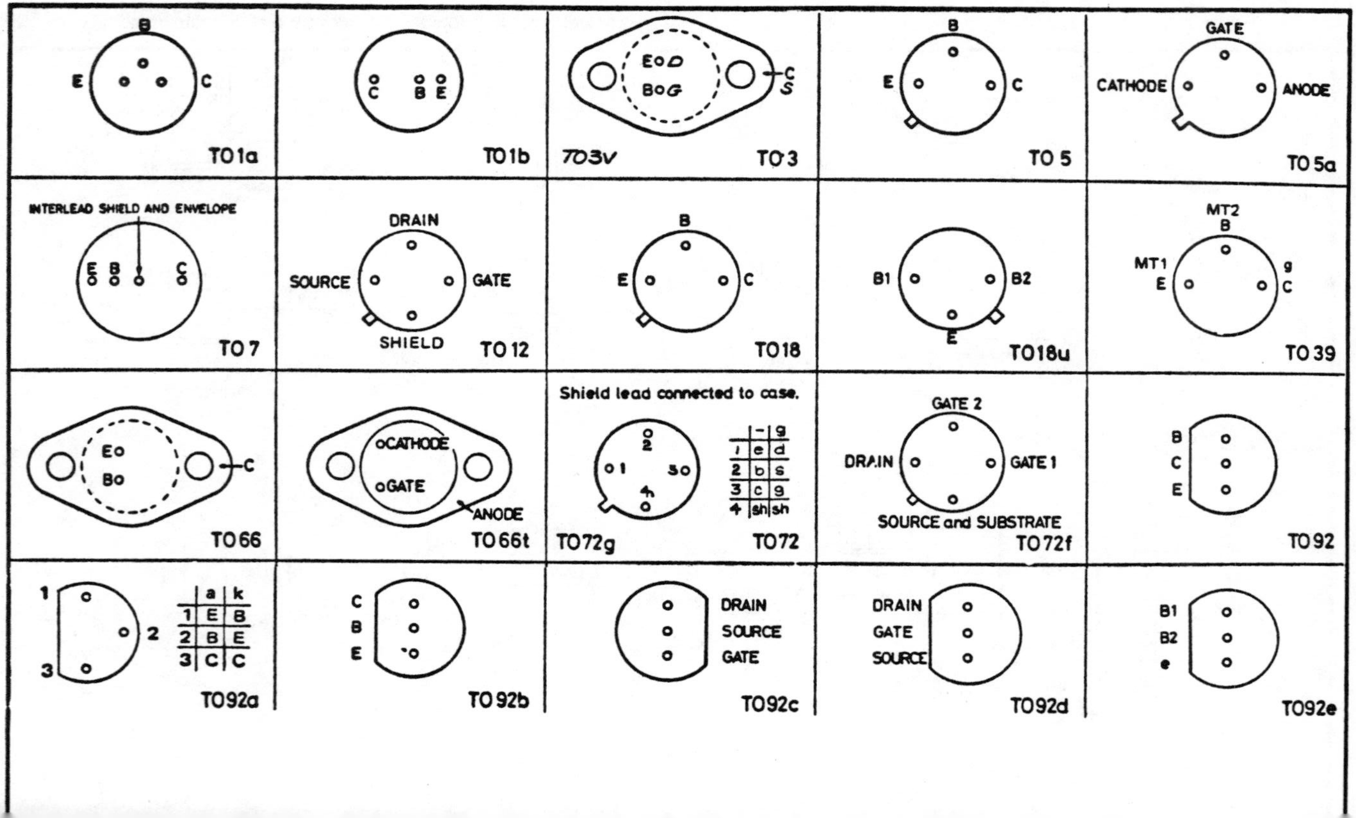

TO5a

GATE   ANODE

CATHODE

TO 39

MT2
B

g
C

MT1
E

TO 92

B C E

TO92e

B1 B2 e

TO 5

B
C

E

TO18u

B2

B1
E

TO72f

GATE 2

GATE 1

DRAIN

SOURCE and SUBSTRATE

TO92d

DRAIN GATE SOURCE

TO3

C
S

E○D
B○G

TO18

B
C

E

TO72

1	e	d	g
1	e	b	s
2	b	c	g
3	c	g	sh
4		sh	

Shield lead connected to case.

3○   ○

○2   f○

○1

TO72g

TO92c

DRAIN SOURCE GATE

TO1b

○ ○
B E
○ ○
C

TO12

DRAIN   GATE

SOURCE   SHIELD

TO66t

○CATHODE   ANODE

○GATE

TO 92b

C B E

TO1a

B
C

E

TO7

INTERLEAD SHIELD AND ENVELOPE

C

E○ B○ ○

TO66

C

E○
B○

TO92a

	a	k
1	E	B
2	B	E
3	C	C

2○

1○   ○

3

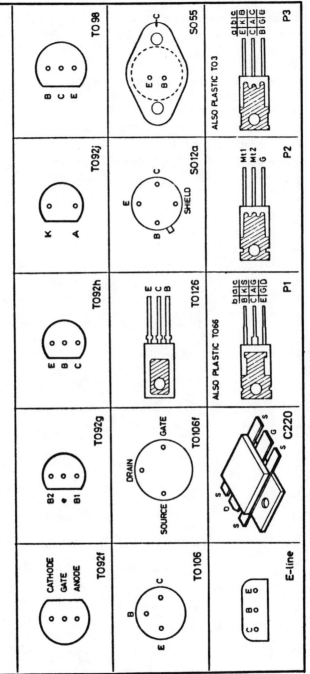

Fig. 7-6. Transistor cases viewed from bottom with lead identification.

circuits (ICs), packaged in the same outline shape as a transistor. More complex ICs are packaged in quite different form—e.g., flat packages.

Power transistors are easily identified by shape. They are metal cased with an elongated bottom with two mounting holes. There will only be two leads (the emitter and base) and these will normally be marked. The collector is connected internally to the can, and so connection to the collector is via one of the mounting bolts or bottom of the can. Examples of transistor outline shapes together with typical dimensions and lead identification are given in Fig. 7-5. Different styles of transistor cases are shown in Fig. 7-6.

Chapter 8

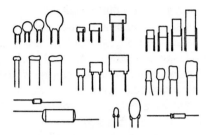

Transistor Characteristics

It is important to have handy for reference (or preferably commit to memory) the symbols used in describing the voltages and currents applicable in basic transistor currents. *Capital* letters are used to designate *average* voltages and currents, and lowercase letters for instantaneous values, i.e., V for average voltages, v for instantaneous voltages, I for average currents, i for instantaneous current.

The elements of the transistor are described in a similar manner, but as subscripts, using B or b for *base*, C or c for collector, and E or e for emitter. For example: V_B, V_c, V_e means average values of base, collector and emitter voltages, respectively, while v_b, v_c, v_e, means the instantaneous values of base, collector and emitter voltages respectively. I_B, I_C, I_E means average values of base, collector and emitter currents, respectively. i_b, i_c, i_e means instantaneous values of base, collector and emitter currents respectively.

Now, besides average values (represented by a capital subscript) and instantaneous values (represented by a lower case subscript), there can also be instantaneous *total* values of voltage or current. These are represented by a capital subscript, but since the value is instantaneous, not average, a lower case letter is used to designate voltage or current. Thus: v_B, v_C, v_E means instantaneous total values of base, collector, and emitter voltage respectively. And i_B, i_C, i_E means instantaneous total values of base, collector and emitter current respectively.

Different voltage and current values (whether average, instantaneous, or instantaneous total) apply in different parts of the circuit. It is often necessary, therefore, to specify the connections of the transistor between which voltages apply. This is simply done by designating those in the subscript symbols, e.g.:

V_{BE} = average base-emitter voltage
v_{be} = instantaneous base-emitter voltage
v_{BE} = instantaneous total base emitter voltage
V_{CE} = average collector-emitter voltage
v_{ce} = instantaneous collector-emitter voltage
v_{CE} = instantaneous total collector-emitter voltage
V_{BC} = average base-collector voltage
v_{bc} = instantaneous base-collector voltage
v_{BC} = instantaneous total base-collector voltage

These subscripts may be used the other way around:

V_{EB} = for average emitter-base voltage
V_{EC} = for average emitter-collector voltage
V_{CB} = for average collector-base voltage

and so on.

Logically they should conform to the voltage direction (positive to negative), depending on whether the transistor is a pnp or npn type, i.e., in the following order:

pnp transistor - EB, BC, BE (or eb, bc, be)
npn transistor - BE, CB, EB (or be, cb, eb)

Other symbols are used to designate the base and supply voltages:

V_{BB} = base circuit bias voltage
V_{CC} = collector circuit supply voltage.

Parameters of importance in determining (or specifying) the performance of transistors (in common emitter configuration—see later) are:

☐ *Input characteristics*—or low base current (i_B) varies with base voltage (v_{BE}).

☐ *Output characteristics*—or low collector current (i_C) varies with bias and collector voltage (or specifically with v_{CE}).

☐ *Transfer characteristics*—or the relationship between collector current (i_C) and base voltage (v_B).

☐ *Current amplification*—or the ratio of collector current (i_C) to base current (i_B) or the signal current *gain*. Alternatively, if the transistor is worked as a *voltage amplifier*, the ratio of the output voltage to the input voltage or voltage gain. This is now designated hf or hfe (alternatively h_{FE}) also known as the β of the transistor.

□ *Power gain*—which is the ratio of output power to input power, determined with respect to current in voltage gain. This is commonly expressed in *decibels* (dB), determined by taking the logarithm of the power gain:

Power gain (dB) = 10 log (output power/input power)

In this case of current amplifications:

Power gain (dB) = 20 log $i2/i_1$ + 10 log R_L/R_1

where, i_1 is the input current
i_2 is the output current
R_1 is the input resistance
R_2 is the output (load) resistance

In the case of voltage amplification the voltage gain is defined as 20 log (v_2/v_1) regardless of the values of R_1 and R_2, v_2 being the output voltage and v_1 the input voltage.

Ratio of collector current to emitter current (or i_C/i_E) is designated as hfb (or hFb).

□ *Quiescent operating or Q-point*—or the actual working point of the transistor. This can be determined by superimposing the *load line* or the characteristic curves of the transistor. It is important that this should be known and stabilized in transistor circuit design (see later).

OUTPUT CHARACTERISTICS

The *output characteristics* (i_C/v_C) of a transistor can be measured in terms of how the collector current (i_C) varies with collector voltage (v_C) for different values of base current (i_B). Typical curves are shown in Fig. 8-1 where it will seem that above the transistor 'knee' point, collector current is fairly constant with increasing collector voltage for all values of bias (base current). It also indicates that the *output resistance* of a transistor is quite low.

INPUT CHARACTERISTICS

The *input characteristic* (i_B/v_B) can be plotted by measuring base current (i_B) against base voltage (v_B) for a constant collector voltage. In this case a typical form of the curve is as shown in Fig. 8-2. The *input resistance* at any point in this case is represented by the gradient of the curve or the ratio of v_B to i_B.

TRANSFER CHARACTERISTICS

The *transfer characteristics* (i_C/v_B) yields a similar shape of

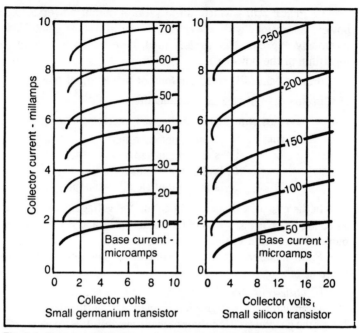

Fig. 8-1. Output characteristics of typical small signal transistors.

curve (Fig. 8-3). These characteristics are determined at a constant collector voltage but since collector current does not vary much with collector voltage, one curve is usually representative of transfer characteristics at all values of collector voltage. This is particu-

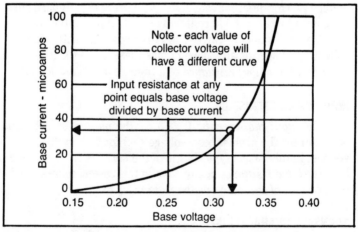

Fig. 8-2. Input characteristics of typical small signal transistor.

larly true of germanium junction transistors. With silicon transistors the characteristic curves tend to depart from the generalized form shown. Also the collector current (leakage current) and zero base current (or zero base voltage) is very much smaller. But the main difference in the characteristics is that a higher base voltage (bias) is needed to cause an appreciable amount of collector current to flow in the case of a silicon transistor, i.e., to operate the transistor above the 'knee' point. This bias is about 0.6-0.7 volts for a silicon transistor but only 0.1-0.2 volts for a germanium transistor.

CURRENT AMPLIFICATION

The *current amplification factor* (i_C/i_B) can be derived by plotting collector current against base current. Typically this gives a straight line relationship (Fig. 8-4). Actual values may range from as low as 10 up to several hundreds, depending on the type of transistor, also the slope of the i_C/i_B curve is not always constant, i.e., the value of *hfe* may have a spread of perhaps 50 percent on either side of a nominal value.

RATIO OF COLLECTOR AND EMITTER CURRENTS

The *ratio of collector and emitter currents* (i_C/i_E) is not a particular significance. Emitter current must equal $i_C + i_B$ but since i_B is

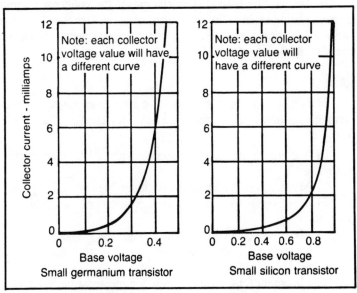

Fig. 8-3. Typical transfer characteristics of small signal transistors.

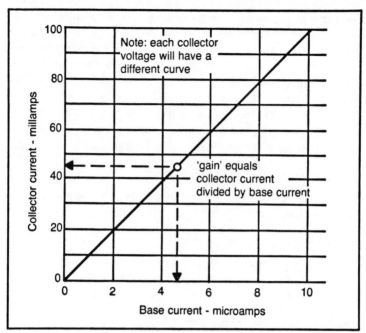

Fig. 8-4. Current amplification produced by a transistor.

small in comparison with i_c, emitter current (i_E) is roughly equal to i_c. The smaller the base current the closer the collector current comes to being equal to the emitter current, and thus the closer the ratio of *hfb* comes to unity. In practice, typical values of *hfb* lie between 0.92 and 0.98.

RATINGS

Individual types of transistors are also given *ratings*, representing maximum values the transistor can handle in a circuit. These are:

☐ Maximum collector-base voltage with emitter open circuit, designated V_{cbo} max.

☐ Maximum collector-emitter voltage, designated V_{ce} max.

☐ Maximum base-emitter reverse voltage or bias, designated V_{eb} max.

☐ Maximum collector current, designated I_c max.

☐ Maximum collector-base current with emitter open circuit, designated I_{cbo} max.

☐ Maximum total power dissipation, designation P_t max.

Chapter 9

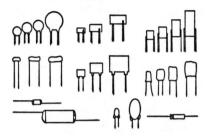

Basic Guide to
Selecting Transistors

Transistors are specified by code letters and/or numbers, which is only a manufacturer's coding. Published circuit designs normally specify a particular type of transistor(s), all the associated component values—e.g., resistors—then being determined with respect to the characteristics of the particular transistor(s) specified. Simply use the specified transistor(s)—unless, as can happen, you find that they are unobtainable.

In that case there are basically three options. The first is to use an *equivalent* transistor of different manufacture in type number which has the same characteristics. For this you need a listing of *transistor equivalents* from which to select an alternative. There are books available which give such equivalent listings. Alternatively, your local Radio Shack should be able to help if the types are not covered in the listing in this chapter.

Equivalents given in such listings are seldom exact equivalents. They are more likely to be near-equivalents with sufficiently close characteristics to be used in 'basic' amounts where component values are not too critical. Simple radio circuits are an example. In many cases with elementary circuits almost any type of transistor of the same basic type (e.g., germanium or silicon), or better still same *functional group*, will work. That is the second option to work with.

Information on functional group is harder to come by. Manufacturers group their products in this way, but suppliers usually only list their stocks by type number, which is not very helpful without manufacturers' catalogs to check on the functional group, to which a particular transistor conforms. Where you can find transistor types listed under functional groups, keep this material on file. It can be an invaluable guide in selecting transistors for a particular job.

Functional grouping normally lists transistor types under specific applications. The following will cover most requirements:

Germanium Transistors
(i)   small, medium currents switching services
(ii)   medium current switching, low power output
(iii) small, medium current amplifiers
(iv) af amplifiers, low power output
(v)   complementary pairs
(vi) high power output (power transistor)

Silicon Transistors
(i)   af amplifiers, small signal, general purpose
(ii)   af amplifiers, low level, low noise
(iii) small signal amplifiers
(iv) rf amplifiers and oscillators
(v)   medium current switching, low power output
(vi) high frequency, medium powers
(vii) general purpose switching
(viii) power transistors

Any reference to 'power' grouping is largely arbitrary since there is no universal agreement on the range of power levels (referring to the maximum power rating of the particular transistor). Thus *low power* may generally be taken to cover 100-250 mW, but such a grouping may include transistors with power ratings up to 1 watt. Similarly *medium power* implies a possible power range of 250 mW to 1 W (but may extend up to 5 watts). Any transistor with a power rating of greater than 5 watts is classified as a power transistor.

PRO-ELECTRON CODING

From the start of production of transistors manufacturers adopted their own individual form of coding, such as a letter identifying the manufacturer or general class of the group (functional group), with a number designating the particular design on development. Several attempts have been made to find a 'universal' code, the most acceptable of which is probably Pro-Electron coding.

This consists of a two letter code followed by a serial number. The first letter indicates the semiconductor material, i.e.,

A = germanium
B = silicon

the second letter indicates the general function of the device, i.e.,

A - detector diode or mixer diode
B - variable capacitance diodes
C - af transistor (excluding power types)
D - af power transistor
F - rf transistor (excluding power types)
L - rf power transistor
S - switching transistor (excluding power types)
V - power type switching transistor
Y - rectifier diode
Z - zener diode

The *serial number* following then consists of three figures if the device is intended for consumer applications, e.g., radio receivers, audio amplifiers, television receivers, etc. A serial number consisting of a single letter followed by two figures indicates the device is intended for industrial or specialized applications. If you find transistors described by this code, then you should have no trouble in deciding to which functional group they belong.

The third option if you cannot obtain a transistor specified for a particular circuit (or where the transistor type is not specified at all) is first to identify which *functional group* it belongs to. If it is used in an audio frequency amplifier, for example, it could be germanium group (iii) or (iv), or silicon (i), (ii) or (iii).

Decide which group logically applies. Then choose an available transistor which belongs in this group and for which the characteristics are available (from the manufacturer's data sheet). Then recalculate the component values required around the characteristics of the chosen transistor (as detailed in the next chapter). Do not assume it will be a 'near-equivalent' to the one used in the original circuit. It may be anything but (especially if the original one specified was a germanium transistor, for example, and you opt to use a silicon transistor as being in the same functional group).

TRANSISTOR EQUIVALENTS

The following listing (Table 9-1) gives approximate equiva-

Table 9-1. Transistor Equivalents.

Type	Equiv.	Type	Equiv.	Type	Equiv.	Type	Equiv.
2G210	OC28	2N256A	AD149	2N576	2N1304	2N1154-56	BC107B
2G301	OC45	2N257	AD149	2N587	AC176	2N1169-70	2N1304
2G302	OC44	2N258-62	2N2906			2N1175	ACY17
2G303	OC45	2N263	2N2907	2N591	AC126	2N1185-90	ACY17
2G304	OC44	2N264	2N2906	2N594	2N1304	2N1191	2N1304
2G371	OC71	2N265	AC128	2N595-96	2N1302	2N1194	ACY17
2G374	OC71	2N266	AC128	2N597	2N1305		
2G381-82	AC128	2N270	AC128	2N609-11	AC128	2N1196-97	BC177
		2N272	AC126	2N613	AC128	2N1198	2N1304
2G395-97	OC70	2N273	AC128	2N619	2N3703	2N1217	2N1304
2N34A	AC126			2N620-21	BC107B	2N1218	AD161
2N35	AC127	2N274	OC170	2N622	BFY50	2N1219-20	BC160
2N35A	AC127	2N279	AC128	2N625	AD161	2N1221-23	2N2904
2N36-38	AC128	2N280	AC126			2N1225	AF115
2N45A	OC72	2N281	AC128	2N631-33	AC128	2N1226	AF114
2N49	AC126	2N285	AD149	2N635-36	2N1304	2N1227	AD149
2N51	AC126	2N291	AC128	2N646	AC128	2N1228-31	2N2904
2N54	ACY17	2N292	2N1302	2N647	AC127		
2N55-56	AC128	2N293	2N1304	2N649	AC127	2N1247-49	BC109C
2N59-60	AC128	2N296	AD149	2N655	ACY17	2N1251	AC127
2N60A	AC126	2N301	AD149	2N656	2N1893	2N1253	BSY95A
2N61	AC128			2N661	AD162	2N1254-59	2N2907
2N64-65	AC126	2N302-03	AC128	2N679	2N1304	2N1263	AC128
2N66-67	AD162	2N319	AC128	2N696	2N1711	2N1273-74	AC128
2N71	AD162	2N320-21	AC126			2N1280	2N1305
2N74-75	AD161	2N322-24	AC128	2N698	2N1711	2N1281-82	2N1307
2N76	AC126	2N325	AD149	2N699	2N1893	2N1284	2N1305
2N79	AC128	2N327-30	2N2906	2N700	AF139	2N1287	AC128
2N86	AC128	2N332	BC140	2N709	2N2369A		
		2N340	2N1893	2N718	2N1711	2N1298	2N1305
2N93	AC126	2N342	2N1893	2N719-20	2N1893	2N1306	2N1304
2N94	2N1304	2N344-45	AC128	2N721	2N2906	2N1308	2N1304
2N95	AC176			2N722	2N2907	2N1314	AD149
2N97	2N1302	2N346	AF114	2N726-27	BC178	2N1320	AC128
2N98	2N1304	2N347-49	2N1893	2N740	BC141	2N1339-42	2N1893
2N100	2N1304	2N350-51	AD149			2N1359	AD149
2N104	AC126	2N352-53	AD149	2N755	2N1893	2N1366	2N1302
2N106-08	AC126	2N354-55	2N3703	2N780	BF115	2N1367	2N1304
2N109	AC128	2N356-57	2N1302	2N800-14	2N1305	2N1378	ACY17
2N113	2N1305	2N362	AC126	2N815-23	2N1302		
		2N363	AC127	2N824-26	2N1305	2N1385	AF139
2N117	BC107B	2N364	2N1304	2N834-35	2N708	2N1388	BFY51
2N118A	2N3055	2N365-66	AC127	2N844-45	2N1893	2N1404	2N1303
2N119-20	2N3055			2N850	BSX20	2N1408	ACY17
2N124	2N1302	2N367-68	AC126	2N852	BSX20	2N1420	2N1711
2N132A	AC128	2N369	AC127	2N858-67	2N2906	2N1429	2N2905
2N145	2N1302	2N372	OC170			2N1431	AC127
2N156	AD149			2N870-71	2N1893	2N1438	AC128
2N160-63	BC107B	2N376	AD149	2N911	BSX21	2N1441-42	2N2906
2N166	2N1304	2N380	AD149	2N913	2N3703	2N1443	2N2907
2N169	AC127	2N386	AD149	2N914	BSX20		
		2N388	AC127	2N915	2N1302	2N1469	2N2906
2N174	OC35	2N389	2N3055	2N923-28	2N2906	2N1478	ACY17
2N176	AD149	2N393	OC170	2N929-30	BC107B	2N1505	BFY51
2N180	AC128	2N399-401	AD149	2N935-46	2N2906	2N1564-65	2N1893
2N181	2N1304			2N947	2N706	2N1566	2N2219
2N183	2N1302	2N405-06	AC126	2N955	BSX20	2N1572-74	2N1893
2N185-86	AC128	2N407-08	AC128			2N1586	2N706A
2N186A	AC128	2N409-10	AF117	2N958-59	2N706	2N16C5	2N1304
2N187-88	AC128	2N413A	OC45	2N981	BC107B	2N1620	2N5458
2N189-90	AC126	2N414	OC45	2N988-89	2N706	2N1623	2N2906
2N191-92	AC128	2N422	AC126	2N995A	BC178		
		2N425-26	2N1305	2N996	2N2906	2N1672	2N1302
2N193-94	2N1302	2N438A	2N1302	2N1000	2N1304	2N1694	2N1302
2N200	AC126	2N439	2N1304	2N1003-04	2N1305	2N1700	BFY50
2N204	AC126	2N440	2N1304	2N1005-06	BC107B	2N1725	2N3055
2N206	AC126			2N1007	AD149	2N1729	2N1303
2N211-12	2N1302	2N445-49	2N1304	2N1008-09	AC128	2N1730	2N1302
2N213-14	AC127	2N451-54	2N3065			2N1731	2N1303
2N215	AC126	2N456A	OC35	2N1010	AC127	2N1732	2N1302
2N216	AC127	2N460-62	AC128	2N1016	2N3055	2N1808	2N1304
2N217	AC126	2N464-67	AC128	2N1024	BC180	2N1837	BF194
2N220	AC126	2N468	AD161	2N1026	BC160		
		2N470	BFY50	2N1027	BC178	2N1838	BF115
2N223	AC128	2N471-72	2N706	2N1038	AD149	2N1839-40	BF194
2N225-27	AC128	2N474	2N706	2N1040	AD162	2N1889	2N1893
2N228-29	AC127	2N476	2N706	2N1051	2N2219	2N1891	2N1304
2N230	AD149			2N1055	2N1893	2N1892	2N1303
2N233	AC126	2N479	2N706	2N1059	AC127	2N1917-18	BC160
2N234A	AD149	2N496	2N3703			2N1923	2N1893
2N235A	AD149	2N497	2N1893	2N1060	2N2219	2N1924	2N1305
2N235B	AD149	2N505	2N1305	2N1086-87	2N1304	2N1944-46	2N1893
2N236	AD149	2N507	AC127	2N1090-91	2N1304	2N1962-63	2N2369A
2N238	AC126	2N508	AC126	2N1092	BFY50		
		2N515-16	2N1304	2N1093	2N1305	2N1972-73	2N2219
2N241A	AC128	2N523-24	2N1305	2N1097-98	AC128	2N1974-75	2N1893
2N242	AC128	2N529-30	2N1303	2N1101-02	AC127	2N1983	2N2219
2N244	BFY50	2N531-33	2N1305	2N1105	2N1893	2N1986	2N2219
2N249	AC128			2N1114	2N1304	2N1988-89	2N1893
2N250	AD149	2N542-43	BFY50	2N1117	2N1893	2N1990	BSX21
2N250A	AD149	2N545	2N1893			2N1991	2N2904
2N253-54	2N1302	2N549-52	2N1893	2N1126-27	ACY17	2N1992	2N1302
2N255	AD149	2N554-55	AD149	2N1128	AC128	2N201/	2N1893
2N255A	AD149	2N556	2N1302	2N1131-32	2N2904	2N2038	2N3053
2N256	AD149	2N557	2N1304	2N1144-45	AC128		
		2N563-72	AC128				
		2N573	AC126				

Table 9-1. Transistor Equivalents. (Continued from page 88.)

Type	Equivalent
2N2039	2N1893
2N2040	2N3053
2N2041	2N1893
2N2042-43	ACY17
2N2062-64	AD149
2N2085	2N1304
2N2106-07	2N1893
2N2108	2N2219
2N2138	AD149
2N2143	AD149
2N2195	BC142
2N2198	2N1893
2N2217	2N2219
2N2218	BFY50
2N2221A	BFY50
2N2222	BFY50
2N2224	2N2219
2N2236	BC142
2N2241	2N2219
2N2243	2N2219
2N2256-57	BSX20
2N2270	2N2219
2N2271	AC128
2N2303	2N2906
2N2309	2N697
2N2310	2N1893
2N2312	2N1893
2N2316	2N1893
2N2360	AF139
2N2380	2N1893
2N2390	2N1711
2N2393-94	2N2905
2N2395-96	2N2219
2N2405	2N1893
2N2410	2N1711
2N2430	2N1304
2N2431	AC128
2N2434	2N1711
2N2437-39	2N1893
2N2443	2N1893
2N2483	BC107B
2N2538	2N2219
2N2570	BSX20
2N2604	BC177
2N2613	AC126
2N2692-93	BCY70
2N2695-96	2N2906
2N2707	AC128
2N2709	2N2906
2N2711-13	BC168C
2N2714	2N708
2N2784	2N2369A
2N2800	2N2904
2N2836	AD149
2N2837-38	2N2904
2N2863-64	2N2219
2N2865	BF180
2N2891	BD140
2N2922	BC209C
2N2923	BC168C
2N2924-25	BC107B
2N2927	2N2904
2N2938	2N2369A
2N2940	BC160
2N2944-45	BC178
2N2953	AC128
2N2958	2N2219
2N2959	2N697
2N2960-61	2N2219
2N3010	2N2369A
2N3020	2N1893
2N3037-38	2N1893
2N3069-71	MEF4220
2N3072	2N2904
2N3073	2N2906
2N3077	2N2484
2N3109	2N697
2N3120	2N2904
2N3121	2N2906
2N3122-23	2N2906
2N3133	2N2904
2N3134	2N2905
2N3135-36	2N2906
2N3224	2N2905
2N3227	2N2369A
2N3232	2N3055

Type	Equivalent
2N3235	2N3055
2N3246-47	2N2484
2N3261	BSX20
2N3279-86	AF239
2N3289	BF200
2N3291-94	BF200
2N3295	2N2219
2N3300	PN3643
2N3304	BSX20
2N3390	BC168C
2N3391	BC107B
2N3392	BC109C
2N3393	BC107B
2N3394-98	BC168C
2N3399	AF139
2N3402-03	BC108C
2N3404	BC337
2N3405	BFY50
2N3414-15	BC338
2N3416	BC182L
2N3417	BC337
2N3467-68	2N2905
2N3485-86	2N2907
2N3498	2N697
2N3499	2N1893
2N3502-03	2N2905
2N3504-05	2N2907
2N3508-09	BSX20
2N3547	BC177
2N3549-50	BC179
2N3566	BC107B
2N3568	BC337
2N3569	BFY50
2N3638	MPS3638
2N3638A	MPS3638A
2N3642	BC337
2N3643	PN3643
2N3644	BC327
2N3646	BSX20
2N3662	BF200
2N3671-73	2N2905
2N3721	BC168C
2N3764-65	2N2907
2N3793	BC168C
2N3798	BC179
2N3824	BC107B
2N3825	BC108C
2N3826-27	2N3903
2N3828	BC109C
2N3831	BD131
2N3838	BC107B
2N3845	BC109C
2N3855	2N3903
2N3856	2N3904
2N3878	BD124
2N3914	2N2906
2N3923	BF337
2N3945	BFY50
2N3946-47	2N2219
2N4014	2N2219
2N4026	BC160
2N4027	2N2906
2N4028-29	2N2907
2N4030-33	2N2905
2N4036-37	2N2905
2N4077	AD161
2N4078	AD162
2N4104	2N2484
2N4121	2N3905
2N4122	2N3906
2N4123-26	2N3905
2N4140-41	BC338
2N4142-43	BC338
2N4227	BC338
2N4228	BC328
2N4237-38	BFY50
2N4239	2N1893
2N4241	AD149
2N4248-49	2N4058
2N4256	2N3904
2N4264-66	BSX20
2N4275	BSX20
2N4288	BC213L
2N4289	BC177
2N4290-91	BC179
2N4293	BC179
2N4294-95	BSX20
2N4348	2N3773

Type	Equivalent
2N4401	PN3643
2N4402-03	2N2904
2N4424-25	BC337
2N4434-35	BC107B
2N4449	BSX20
2N4452	MPS3638
2N4890	2N2904
2N4918	BD132
2N4920	BD140
2N4921	BD131
2N4923	BD139
2N4924-25	2N1893
2N4926-27	BF337
2N4944	BC338
2N4945	BC337
2N4946	BC338
2N4951-53	BC337
2N4954	BC338
2N4964	2N4060
2N4965	2N4058
2N4967	BC209C
2N4969-70	BC337
2N4971-72	BC328
2N4994	2N3903
2N5030	BSX20
2N5036	2N3055
2N5037	BD131
2N5058	BF259
2N5083	2N3055
2N5086	BC177
2N5087	BC179
2N5088-89	BC169C
2N5106	2N2219
2N5128-29	BC338
2N5133	BC209C
2N5134	BSX20
2N5137	BC338
2N5138	2N4061
2N5163	2N3819
2N5189	2N3053
2N5191-92	TIP31A
2N5194-95	TIP32A
2N5208	2N3905
2N5220-21	BC338
2N5223	2N3903
2N5224	BSX20
2N5225-26	BC338
2N5293-94	TIP31A
2N5297-98	TIP31A
2N5306	MPSA14
2N5308	MPSA14
2N5354-56	BC328
2N5365	BC327
2N5366	BC337
2N5367	BC327
2N5380	2N3903
2N5381	2N3904
2N5382	2N3905
2N5383	2N3906
2N5447	BC328
2N5448	MPS3638
2N5449-51	BC337
2N5758-60	2N3773
2N5814	BC140
2N5815	BC327
2N5816	BC140
2N5817	BC337
2N5818	BC140
2N5819	BC327
2N5825	BC140
2N5827	BC209C
2N5858	2N1893
2N5881-82	2N3055
2N5961	BC107B
2N6003	BC179
2N6008	BC109C
2N6015	BC177
2N6064-67	2N2907
2S22	AC128
2S37-40	AC128
2S41	AD149
2S43-44	AC128
2S54	AC126
2S56	AC128
2S501	BC109C

Type	Equivalent
2S502-03	BC107B
2SA12	OC44
2SA17	OC44
2SA28	OC44
2SA32-33	OC44
2SA64	OC44
2SA79	2N1305
2SA127	OC44
2SA128-29	2N1305
2SA169-71	OC44
2SA172-74	2N1305
2SA189	OC44
2SA204-07	2N1305
2SA217	2N1305
2SA229-30	AF139
2SA242-45	AF239
2SA248	2N1305
2SA264-66	AF239
2SA277-78	2N1305
2SA282-84	2N1305
2SA288-90	AF239
2SA296-97	OC44
2SA311-12	2N1305
2SA325-26	2N1305
2SA373	AF239
2SA378	AF239
2SA385	OC44
2SA413	AF239
2SA414-15	2N1305
2SA417	AF239
2SA419	AF239
2SA420	AF139
2SA421	AF239
2SA422	AF139
2SA434-38	AF239
2SA454-56	AF139
2SA458	2N1305
2SA478-79	2N1305
2SA495	2N4058
2SA496	BD140
2SA499-500	BC177
2SA527	2N3703
2SA537	BFX29
2SA538	OC44
2SA539	2N4061
2SA550	BC179
2SA561	BC213L
2SA564	2N4061
2SA565	BFX29
2SA567	BC179
2SA571	BFX29
2SA594	BFX29
2SA628-29	2N4061
2SB44	AC126
2SB46-50	AC126
2SB51-54	AC128
2SB56	AC128
2SB57	AC126
2SB59-60	AC126
2SB63	AD149
2SB66	AC126
2SB73-76	AC126
2SB77-78	AC128
2SB80	AD149
2SB83	AD149
2SB87	OC28
2SB89	AC128
2SB90	AC126
2SB91	AC126
2SB94	AC128
2SB97	AC126
2SB99	AC126
2SB102-04	AC128
2SB105	AD149
2SB108-09	AD149
2SB111	AC126
2SB114-17	AC128
2SB131	AD149
2SB134-35	AC126
2SB136	AC126
2SB137	AD149
2SB140	AD149
2SB142-46	AD149
2SB153	AC126
2SB154-56	AC128
2SB157-60	AC126
2SB161-67	AC128

Table 9-1. Transistor Equivalents. (Continued from page 89.)

Type	Equiv.
2SB168	AC126
2SB169	AC128
2SB172	AC128
2SB173	AC126
2SB174	AC128
2SB176-77	AC128
2SB181	AD149
2SB183	AC126
2SB184	AC128
2SB185-87	AC126
2SB188-89	AC128
2SB199	AC128
2SB200	AC128
2SB202	AC128
2SB208	AC128
2SB216	AD149
2SB219-23	AC128
2SB249	AC128
2SB253-55	AC128
2SB256	AD162
2SB261-62	AC126
2SB263	AC128
2SB264	AC126
2SB265	AC128
2SB293-94	AC128
2SB328-29	AC128
2SB345-48	AC126
2SB350	AC128
2SB364	AC128
2SB370	AC128
2SB376	AC128
2SB378	AC128
2SB381	AC326
2SB386	AC128
2SB431	AC128
2SB439	2N2907
2SB440	AC128
2SB457	AC128
2SB461	ACY17
2SB463	AD162
2SB473	AD162
2SB475	AC126
2SB476	ACY17
2SB492	ACY17
2SB534	AC128
2SC11	2N1304
2SC13-14	2N1304
2SC16-18	BC107B
2SC21	2N3055
2SC30	2N1711
2SC31	BC140
2SC32	BFY50
2SC34-36	2N1304
2SC37-38	2N1711
2SC39	BC107B
2SC45	BFY50
2SC59	2N1893
2SC60	2N1304
2SC61	BFY50
2SC69	2N1893
2SC74	BFY50
2SC89-91	2N1304
2SC97	BC140
2SC103	BC107B
2SC105	BC169C
2SC108-09	BFY50
2SC120	BC140
2SC128	2N1302
2SC129	2N1304
2SC170	BC108C
2SC178	2N1304
2SC183-84	BF115
2SC186-87	BC107B
2SC191-97	BC107B
2SC206	BF184
2SC233	2N1893
2SC248	BC107B
2SC281	BC107B
2SC282	BC108C
2SC283	2N2484
2SC285	2N2219
2SC297	2N3054
2SC299	2N3054
2SC306-07	BFY50
2SC310	2N1893
2SC313	BC108C
2SC316	BC109C
2SC318	BC107B

Type	Equiv.
2SC320	2N2219
2SC350	BC107B
2SC352	2N7219
2SC354	BFY50
2SC360	BC107B
2SC363	2N2219
2SC366	2N3704
2SC368	BC107B
2SC369-74	BC168C
2SC379	BC169C
2SC381	BC168C
2SC400	BC107B
2SC405-06	2N1304
2SC454	BC148
2SC458	BC148
2SC460-61	BC148
2SC475-76	BC179
2SC477-78	BC107B
2SC482	BFY50
2SC484-86	2N1893
2SC490-91	2N3054
2SC495-96	BD139
2SC497	2N1893
2SC498	BFY50
2SC503	BFY50
2SC505-06	BF259
2SC510-12	2N1893
2SC513	BFY50
2SC516	2N1893
2SC536	BC108C
2SC561	BC108C
2SC589	BF259
2SC594	2N2219
2SC596	2N2219
2SC620	BC337
2SC622	BC108C
2SC627	BF259
2SC631	BC107B
2SC632	BC108C
2SC633-34	BC107B
2SC640	BC108C
2SC649-50	BC107B
2SC656	2N3904
2SC686	BF167
2SC708	2N2219
2SC712	BC168C
2SC714	BC337
2SC727	2N1893
2SC732	BC184L
2SC733	BC182L
2SC734	BC107B
2SC736	2N3704
2SC752	BSX20
2SC756	2N1893
2SC763	2N3704
2SC776	2N2219
2SC778	2N3064
2SC784	BF200
2SC788	BF259
2SC802-03	BFY50
2SC815	2N3704
2SC816	BFY50
2SC838	BC149
2SC856	BSX21
2SC864	BF167
2SC868-69	2N1893
2SC870-71	BC209C
2SC894	BC108C
2SC899	BC149
2SC912	BC108C
2SC926	2N1893
2SC941	BC168C
2SC943	BC107B
2SC983	2N1893
2SC995	BF258
2SC997	BF167
2SC1000	BC209C
2SC1033	2N1893
2SD30-38	AC127
2SD43-44	AC127
2SD46	2N3055
2SD51	2N3055
2SD53-54	2N3055
2SD59-60	2N3065
2SD61-88	AC127
2SD70	BD124
2SD73	2N3055
2SD75	AC127
2SD77-78	AC127
2SD80-83	2N3055

Type	Equiv.
2SD96	AC187
2SD104-06	AC127
2SD124-25	2N3065
2SD144	2N3064
2SD146	2N3064
2SD151	2N3056
2SD172-73	2N3056
2SD175-76	2N3055
2SD180	2N3055
2SD186-87	AC127
2SD196	AC127
2T64-67	AC127
2T69	AC127
2T85	AC127
2T311-15	AC128
2T321-24	AC128
2T681	AC127
2T3081-33	AD149
2T0041-43	AD149
3N159	3N140
4J01A	AC128
40217-19	BC107B
40221-22	BSX20
40231-32	BC107B
40234	BC107B
40242	BF200
40250	2N3064
40251	2N3065
40253	AC128
40255	BF258
40264	BF337
40269	ACY17
40310	2N3054
40311	BFY50
40315	BFY50
40317	BFY50
40320	BFY50
40325	2N3055
40326	BFY50
40361	BFY50
40363	2N3055
40364	TIP31A
40372	2N3054
40389	2N3053
40539	BFY50
4057B	2N3866
40611	BFY50
40635	2N3053
40636	2N3055
AC105	AC128
AC110	AC128
AC114-15	AC128
AC120-21	AC126
AC122	AC126
AC130	AC176
AC134	OC71
AC139	AC128
AC150-51	AC128
AC152	AC128
AC154	AC128
AC157	AC127
AC160	AC127
AC162	AC126
AC163	AC128
AC166-67	AC128
AC168	AC127
AC171	AC126
AC172	AC127
AC173	AC128
AC175	AC127
AC179	AC176
AC180	AC128
AC181	AC176
AC182	AC128
AC183	AC127
AC185	AC127
AC186	AC176
AC192	AC126
AC193	AC128
ACY18	AC128
ACY27-29	OC70
ACY39	AC128
AD130-31	AD149
AD132	OC28
AD133	AD149
AD138	OC28

Type	Equiv.
AD142-43	OC28
AD148	AD162
AD150	AD149
AD153	AD162
AD155	AD162
AD157	AD161
AD164	AD162
AD165	AD161
AD166-67	AD149
AD169	AD162
AD262	AD162
AF147	AC127
AF186	AF139
AF192	AC127
AF240	AF139
AF250-52	AF239
AF253	AF139
AF254	AF239
AFY18	AF139
AFY19	BFX88
ASY24	AF239
ASY26-27	BFX29
ASY28-29	BFY50
ASY50	AC128
ASY54-56	2N1303
ASY57	2N1305
ASY67	BCY70
ASY70	AC128
ASZ20	BCY70
ASZ23	BCY70
AT318-19	BF115
AU102	AC128
AUY33	AD149
BC113	BC149
BC114	BC149
BC116	BC117
BC125	BC182L
BC126	BC143
BC132	BC108C
BC134	BC182L
BC144	BC142
BC145	BC117
BC152	BC337
BC167	BC182L
BC170	BC148
BC171	BC107B
BC175	BC337
BC180	BC140
BC181	BC204
BC185	2N2219
BC186	BC178
BC187	BC179
BC190	2N3904
BC192	2N2907
BC194	BF115
BC196	BF167
BC199	BC109C
BC201	BC109C
BC205	BC158
BC206	BC159
BC207	BC107B
BC208	BC108C
BC210	BC337
BC211	BFY50
BC215	BC327
BC216	BC328
BC218	BC107B
BC219	BC140
BC220	BC148
BC221	MPS3638A
BC222	BC337
BC223	BC140
BC224	BC158
BC225	BC157
BC226	BC140
BC231	BC177
BC232	BC107B
BC234-35	BC107B
BC237	BC107B
BC238	BC108C
BC239	BC109C
BC250	BC177
BC251	BC157
BC252	BC158
BC253	BC159
BC256-57	BC157
BC258	BC212L
BC259	BC159

Table 9-1. Transistor Equivalents. (Continued from page 90.)

Type	Equiv.
BC260	BC158
BC262	BC158
BC263	BC159
BC264	BCY71
BC266	BC212L
BC270	BFY50
BC271	BC338
BC272	BC337
BC274	BC157
BC275	BC158
BC276	BC159
BC277	BC107B
BC278	BC108C
BC279	BC109C
BC280	BC140
BC281	BCY70
BC282	BC337
BC283	BC327
BC284	BC107B
BC286	BC141
BC287	BC160
BC288	BC141
BC289	BC107B
BC290	BC184L
BC291	BC157
BC292	BC159
BC293	BFY51
BC294	BC161
BC295	BC107B
BC297-98	BC160
BC300	BC141
BC302	BC140
BC304	BC461
BC307	BC177
BC308	BC178
BC309	BC179
BC310	BC141
BC311	BC160
BC312	2N1893
BC313	BC161
BC315	BC179
BC319	BC109C
BC320	BC327
BC323-24	BFY50
BC325-26	BCY71
BC332	BC107B
BC333	BC108C
BC340	BC337
BC341-42	BFY50
BC344	BFY50
BC349	BC169C
BC360	BC327
BC361	BC157
BC362-64	BD140
BC370	BC160
BC371	BFY51
BC377	BC337
BC378	BC338
BC383	BC337
BC384	BC109C
BC385	BC107B
BC386	BC106C
BC387	BC337
BC388	BC327
BC389-91	BC109C
BC405	BC212L
BC406	BC177
BC408	BC108C
BC409	BC209C
BC411-12	BC141
BC415A	BC214L
BC416A	BC214L
BC417	BC157
BC418	BC158
BC419	BC159
BC430	BD140
BC440	BD140
BC460	BC160
BC478	BC178
BC479	BC179
BC491	BC461
BC507	BC182L
BC508	BC183L
BC509-10	BC184L
BC512	BCY70
BC513	BC158
BC514	BC159

Type	Equiv.
BC549	BC338
BC582	BC107B
BC583	BC108C
BC584	BC109C
BCW10	BC338
BCW11	2TX500
BCW12-13	BC327
BCW14	BC337
BCW15	BC327
BCW16	BC337
BCW17	BC327
BCW18	ZTX304
BCW19	ZTX504
BCW20	BC337
BCW21	BC327
BCW22	ZTX331
BCW23	BC327
BCW25	BC337
BCW29	2N4060
BCW32	2N2484
BCW44-45	2N1893
BCW47	BC147
BCW48	BC148
BCW49	BC149
BCW50	ZTX304
BCW57	BC157
BCW58	BC158
BCW59	BC159
BCW77-78	BFY51
BCW90A	BC328
BCW91A	BC338
BCW92A	BC328
BCW93A	BC327
BCW94A	BC337
BCW95A	BC337
BCW96A	BC327
BCW97A	BC327
BCX31	BFX84
BCX32	BFY50
BCX34	BFY52
BCX35-36	BFX87
BCX37	BFX88
BCX40	BFX85
BCY13-16	2N2219
BCY17-21	2N2906
BCY23-34	2N2906
BCY38-40	2N2906
BCY42-43	BC107B
BCY49	2N2906
BCY50-51	BC108C
BCY54	BCY70
BCY57	BC182L
BCY58	BC107B
BCY59	BC109C
BCY65	BC107B
BCY67	BCY71
BCY90	2N2906
BCZ11	BC177
BCZ14	BC178
BD101-03	MJ2955
BD109	2N3054
BD111	2N3054
BD116-18	2N3055
BD121	2N3055
BD123	2N3055
BD128-29	BF259
BD130	2N3055
BD142	2N3055
BD144	MJE340
BD148-49	2N3084
BD161-63	2N3054
BD165	BD135
BD169	BD139
BD170	BD140
BD181-83	2N3055
BD214	2N3064
BD222	TIP31A
BD232	MJE340
BD605	2N2907
BD702	TIP32A
BDX10	2N3055
BDX18	MJ2955
BDX23	2N3055
BDX24	2N3054
BDX35-37	TIP31A
BDX51	2N3773
BDY13	BD124
BDY15A	TIP31A

Type	Equiv.
BDY16A	TIP31A
BDY17-20	2N3055
BDY26	2N3055
BDY34	TIP31A
BDY39	2N3055
BDY55-56	2N3055
BDY62	2N3054
BDY69	2N3054
BDY71	2N3054
BDY73	2N3055
BDY78-79	2N3054
BF119	BF258
BF121	BF196
BF127	BF167
BF134	BF115
BF137	BF115
BF163	BF196
BF168	BF167
BF178	BF258
BF179	BF259
BF188	BF167
BF189	BF115
BF198	BF167
BF207	BF167
BF214	BF194
BF216	BF115
BF220	BF200
BF222	BF195
BF228	BSX21
BF230	BF195
BF245	BF244
BF250	BC109C
BF251	BF167
BF254	BF194
BF255	BF195
BF257	BF258
BF260	BF167
BF265	BF180
BF267	BF167
BF270	BF167
BF275	BF200
BF287	BF167
BF297	BF258
BF299	BF259
BF314	BF200
BF322	BF258
BF323	BC327
BF324	BF200
BF375	BF167
BF332	BF194
BF333	BF195
BF337	BF258
BF338	BF259
BF367	BF167
BF381	BF258
BF390	BF259
BF450	BF167
BF458	BF258
BF459	BF259
BF594	BF194
BF595	BF195
BFR22	2N1893
BFR23-24	2N2904
BFR50	BFY51
BFR52	BFY50
BFR99	BC327
BFS10	2N3866
BFS72	2N3823
BFS98	2N2904
BFS99	BF337
BFT48-49	BF258
BFV64	2N2907
BFV68	BC107B
BFV82	2N706
BFV83	2N708
BFV86	2N2907
BFW20-23	2N2907
BFW25	2N697
BFW29	BC377
BFW31	2N2907
BFW33	2N1893
BFW34	BFY50
BFW35	BFY51
BFW45	BSX21
BFW61	2N3819
BFW73-74	2N3866
BFW76	2N3866
BFW78	2N3866

Type	Equiv.
BFW80	BFY51
BFW89	BFX87
BFX37	2N2906
BFX38-41	BC161
BFX43	BC108C
BFX48	2N2905
BFX55	2N3866
BFX61	2N1893
BFX62	BF180
BFX65	BC179
BFX68	2N1711
BFX69	BC140
BFX74	BC160
BFX77	BF115
BFX86	BFY50
BFX93A	2N2484
BFX94-96	BC140
BFX97	2N2219
BFY12-13	2N2219
BFY17-19	2N708
BFY26	2N708
BFY28	2N708
BFY33	BC142
BFY40	BFY50
BFY41	2N1893
BFY46	2N1711
BFY49	BC107
BFY55	BFY50
BFY64	2N2905
BFY66	BF180
BFY67	BFY50
BFY68	2N1711
BFY69	2N706A
BFY72	2N2219
BFY76	BC107B
BFY99	2N706A
BLY47A	2N3054
BLY48	2N3055
BSS10	BSX20
BSS11	2N2369A
BSS12	BSX20
BSS13	2N3053
BSS17-18	2N2904
BSS30	2N1893
BSS32	2N1893
BSV23	BSX20
BSV24	BC337
BSV25	2N706A
BSV26	BSX20
BSV27	2N2369A
BSV29	ZTX342
BSV33	BSX20
BSV51	BSX21
BSV84	BFX85
BSV89	2N706A
BSV90	2N708
BSV91	2N706A
BSV92	2N708
BSW21	2N2906
BSW22	2N2907
BSW23	2N2904
BSW24	2N2906
BSW25	BCY72
BSW32	BSX21
BSW36	2N2905
BSW41	2N2907
BSW42-43	BC338
BSW44-45	BC328
BSW52	2N2219
BSW70	BSX21
BSW72-73	BC327
BSW74	2N2906
BSW75	2N2907
BSX27	BSX20
BSX28	2N706A
BSX29	2N2906
BSX35	BSX20
BSX36	2N2906
BSX38	2N708
BSX39	BSX20
BSX40	2N2904
BSX41	2N2905
BSX54	2N1711
BSX63	2N1893
BSX66-67	BC108C
BSX68-69	BC148
BSX70-71	2N2219
BSX74	2N2219

Table 9-1. Transistor Equivalents. (Continued from page 91.)

Type	Equivalent
BSX76-78	2N2369A
BSX79	2N2219
BSX80	2N708
BSX87	BSX20
BSX88	2N708
BSX89	2N706A
BSX93	BSX20
BSY19	2N708
BSY22	BF115
BSY25	2N2219
BSY26	2N706A
BSY28-29	2N708
BSY38-39	2N708
BSY45	2N1893
BSY47-48	2N706A
BSY50-51	2N697
BSY52	2N1711
BSY54	2N1711
BSY55-56	2N1893
BSY59	BC328
BSY61-62	2N706A
BSY63	2N708
BSY70	2N706
BSY71	2N1711
BSY73	2N2369A
BSY77	2N2219
BSY82	2N2219
BSY85-88	2N1893
BSY90	2N1711
BSY91	BC142
BSY92	2N2219
BU104	R2008B
BU105	BU205
BU108	BU208
BU111	BU204
BU126	BU204
BU207	BU208
BUY16	2N3055
BUY38	2N3054
BUY46	2N3054
C420	BFY51
C424	2N3866
C442	2N3866
D29F1	2N3905
D29F2	2N3905
D29F3	2N3906
D43C5	TIP32A
DW6208	BC107B
DW6577	BC107B
DW6737	BC107B
DW7000	BF167
E100	2N3819
E103	2N3819
E300	2N3819
EN697	2N697
EN706	2N706
EN708	2N708
EN722	BC328
EN744	BSX20
EN870-71	2N1893
EN914	BSX20
EN956	BC338
EN1613	BFY50
EN1711	2N1711
EN2219	2N2219
EN2369A	2N2369A
EN2484	2N2484
EN2905	2N2905
EN2907	2N2907
EN3009	BSX20
EN3011	2N2369A
EN3013	BC107B
EN3250	BFX88
EN3502	2N2905
EN3504	2N2907
EN3904	2N3904
EN3905	2N3905
EN3906	2N3906
FE5458	2N5458
FE5459	2N5459
FM1711	2N1711
FPS6514-15	BC169C
FPS6516-18	BC327
FPS6533-35	BC327
FPS6562-63	BC327
FPS6565	2N2904
FT1746	BC177
FT3643	2N2219

Type	Equivalent
FT5040-41	2N3905
GET881-82	2N1305
GET895	2N1305
GET897-98	OC44
GET914	2N2369A
GET2369	2N2369A
GET3013-14	2N2369A
GET3638	MPS3638
GET3638-13	MPS3638A
GET5306	MPSA14
GET5308	MPSA14
GFT2C06/30	AD149
GI2923-26	BC108C
GI3638	MPS3638
GI3644	2N2905
GI3702-03	BC327
GI3704-06	BC140
GI3707-11	BC301/5
GMO290	AF139
GT20	AC128
GT46-47	AC126
GT74	AC126
GT81	AC126
GT109	AC128
GT122	AC126
GT222	OC71
KM7000	AD149
KM7007	AD149
LDA404	2N697
LDA405	2N1711
LDS201	2N2369A
LDS205	2N2369A
MAO401-02	BPX29
MAO404	2N2906
MAO411-14	2N3703
MA9001-03	2N2369A
ME0404	BC328
ME0404/1	BC327
ME0404/2	BC327
ME0411	BC204
ME0414	BC157
ME0463	2N3703
ME0475	2N3702
ME0501-03	BC160
ME0511-12	BC327
ME4001-02	2N3706
ME4101	2N3705
ME4102-03	2N3706
ME4104	BC107B
ME6001-02	2N3706
ME6101	2N697
ME6102	2N1711
ME8002	2N1893
ME8003	2N1711
ME9003	BSX20

Type	Equivalent
ME9022	2N3706
MJ3202	MJE340
MJE105	BD132
MJE170-72	BD140
MJE180-82	BD139
MJE205	BD131
MJE720-22	BD139
MJE2371	TIP32A
MJE2521	TIP31A
MJE2901	MJE2955
MM1755	2N697
MM1756	PN3643
MM1758	2N1711
MP8113	BD139
MP8121-23	BD139
MP8213	2N3054
MP8223	2N3054
MP8511-13	BD140
MP8521-23	BD140
MPF102	2N3819
MPF104	2N5458
MPF105	2N5459
MPS404	2N3904
MPS3390-91	BC169C
MPS3398	BC168C
MPS3693	2N3903
MPS3694	2N3904
MPS3702	MPS3638A
MPS3703	MPS3638
MPS3704	2N3704
MPS3705	2N3705
MPS3706	2N3706
MPS3707	2N3707
MPS3708	2N3708
MPS3709	2N3709
MPS3710	2N3710
MPS3711	2N3711
MPS3827	BC169C
MPS4354-56	2N3903
MPS5172	2N3903
MPS6533-35	BC177
MPS6564	2N3903
MPSA14	2N3903
MPSA16	BC169C
MPSA18	BC169C
MPSA20	2N3903
MPSA66	MPSA65
MPSA70	2N3905
MPSU06	BD139
MPSU51-52	BD140
MPU131	BRY39
MU4891-92	2N2646
MU4893-94	2N2647
NKT4	OC44
NKT5	OC44
NKT32-33	OC44
NKT42-43	OC44
NKT52-54	OC44
NKT74	OC44
NKT125	2N1305
NKT128	AC128
NKT131-32	OC44
NKT133	2N1305
NKT151	AC128
NKT202-03	AC128
NKT210	AC128
NKT211	OC84
NKT212	OC72
NKT213	OC75
NKT214	OC71
NKT215-16	AC128
NKT218	AC128
NKT222-25	AC128
NKT229	AC128
NKT237	ACY17
NKT239	ACY19
NKT240	ACY20
NKT241	ACY21
NKT242	ACY22
NKT244	OC70
NKT261-64	AC128
NKT265	OC70
NKT270	AC128
NKT271-72	OC81D
NKT273-74	AC128
NKT275	OC81D
NKT278	AC128
NKT302	ACY17

Type	Equivalent
NKT304	AD149
NKT351	AC128
NKT402	OC28
NKT403	AD140
NKT450	AD149
NKT451-53	AD149
NKT675	OC170
NKT713	AC127
NKT717	AC127
NKT734	AC127
NKT736	AC127
NKT773	AC176
NKT781	AC127
NKT10339	BC337
NKT10419	BC168C
NKT10439	BC337
NKT10519	BC168C
NKT12329	BC338
NKT12429	BC338
NKT20329	BC328
NKT20339	BC327
OC16	AD162
OC24	AD149
OC26	AD162
OC74-76	AC178
OC77	ACY21
OC79	AC178
OC80	AC126
OC122-23	BFX87
OC139	2N1302
OC140	2N1304
OC141	AC127
OC468	BC178
OC871/72	2N1303
OC880	2N1305
OCP71	BPX75
OD22	AD149
OD74-75	AD149
OD603	AD149
PET3702	MPS3638A
PET3703	MPS3638
PET3903	2N3903
PET3904	2N3904
PET3905	2N3905
PET3906	2N3906
PET8005-06	2N3903
PET8007	2N3904
PET8350	MPS3638
PET8351	MPS3638A
PET8352	MPS3638
PET8353	MPS3638A
PN3638	MPS3638
S1024	BC177
S2048	BC177
S2049	BC140
S2050	BC107
S2292	BC177
SDT1962	2N4060
SDT3322	TIP32A
SDT3501	TIP32A
SDT3776	TIP32A
SDT4301-02	TIP31A
SDT4611	TIP31A
SDT4614	TIP31A
SDT5097	TIP31A
SE1002	BF194
SE1010	BF194
SE2002	BC147
SE3001-02	BF194
SE4001	BC107B
SE5002-03	BF200
SE5023-24	BF167
SE5055	BF167
SE5056	BF196
SE6001	BC107B
SE6021	2N697
SE6562-63	2N2905
SE7055-56	BF258
SE8042	BFY50
SFT106-08	OC44
SFT124-25	AC128
SFT126-28	OC44
SFT130-31	AC128
SFT141-42	AC128
SFT143	AC128
SFT144-46	AC127
SFT184	AC127
SFT186	BF117

Table 9-1. Transistor Equivalents. (Continued from page 92.)

SFT221	2N1303	TIP29B	2N3055	TR-C71	AC126	ZT781	2N2906
SFT223	AC128	TIP30	TIP32A	TR-C72	AC126	ZT691	2N691
SFT226	2N1305	TIP30A	TIP32A	T2551-52	BC328	ZT706	2N706
SFT232	AC128	TIP31	TIP31A	T2554	BC328	ZT708	2N708
SFT306-08	2N2905	TIP31B	2N3055	T2582	BC327	ZT709	2N2369A
SFT321-23	AC126			W3819	2N3819		
SFT325	AC128	TIP32	TIP32A	XA101	OC45	ZT1702	2N2906
SFT337	AC126	TIP33A	MJE3055	ZT20	BFY50	ZT1708	2N706A
SFT351-53	AC126	TIP33B	2N3055	ZT22-24	BFY50	ZT1711	2N1711
SFT377	AC176	TIP34	MJE2955	ZT40-41	BC108C	ZTX114	BC169C
		TIP35B	2N3055			ZTX212	BC212L
TF65	AC126	TIP41	TIP41A	ZT42-43	BC107B	ZTX213	BC213L
TF70-72	AC128	TIP41B	2N3055	ZT66	2N1893	ZTX214	BC214L
TF75	AC128	TIP2955	MJE2955	ZT68	2N1893	ZTX310	2N706
TF77	AD162	TIP3055	MJE3055	ZT80	AC127	ZTX311	BC337
TF78	AD149	TIS37-38	2N3905	ZT86	BSX21	ZTX350	2N3905
TI156	AD149			ZT88	BSX21		
TI407-09	2N3866	TIS45	2N708	ZT91	2N2219	2G102	2N2907
TI412-13	BC337	TIS52	2N3706	ZT92-93	2N1893	2G108	2N2907
TI539	AD149	TIS60	2N3866	ZT118-19	2N1893		
TI1125-26	2N3055	TIS61	2N2906	ZT132	BSX21		
		TIS91	2N2905				
TI1135-36	2N3055	TIS94	2N2905	ZT152-53	2N708		
TI1145-46	2N3055	TP3638	MPS3638	ZT180-84	2N2906		
TI1155-56	2N3055	TP3638A	MPS3638I	ZT187	2N2906		
		TR722	AC126	ZT189	2N2906		
TIP29	TIP31A	TR-C70	AC126	ZT280-84	2N2906		
TIP29A	TIP31A						

lents for a wide range of transistors. The transistor type given in the *second* column should be a suitable substitute for the one in the *first* column. It does not follow, however, that the transistor type in the *first* column will be a suitable substitute for a transistor listed in the *second* column. In other words, only the *second column transistors* can be regarded as substitutes (for first column transistors).

Chapter 10

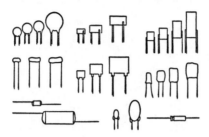

Basic Transistor Circuit Design

Transistors can be operated in any one of three circuit configurations:

☐ Common base (or grounded base)
☐ Common emitter (or grounded emitter)
☐ Common collector (grounded collector)

In all cases the base forms are terminals of the input circuit and the collectors are terminals of the output circuit. Polarities are shown in Fig. 10-1, those of a pnp transistor being opposite to those of an npn transistor. An easily remembered rule in this respect is that the polarity of the *collector* must always be:

> Negative for a p*n*p transistor
> Positive for an n*p*n transistor
> (Remember the middle letter!)

BIAS

The operating characteristics of a transistor are determined by the bias which, in practice, is the voltage applied to, or the current between the emitter and base. This can be referred to as *voltage* bias or *current* bias, respectively. The following basic observations apply:

☐ The forward *current* at the emitter-base junction controls the current between emitter and collector.

☐ An increase in bias current increases the current between emitter and collector; and vice versa.

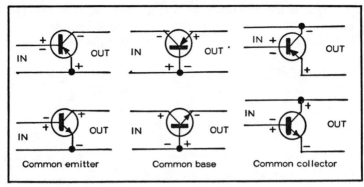

Fig. 10-1. Transistor configurations and polarities.

☐ The emitter current is equal to the sum of the base current and collector current.

COMMON-EMITTER CONFIGURATION

The common-emitter is the most widely used particularly in amplifier circuits. The three elements of the transistor represent separate resistance in a simple direct circuit as shown in Fig. 10-2. The *input* circuit embraces the base resistance (R_B) and the emitter resistance (R_E). The *output* circuit embraces the collector resistance (R_C) and any load resistance (R_L), and the emitter resistance (R_E). In other words the emitter resistance is common to both input and output circuits.

The *input resistance* to a simple direct voltage is represented by the base resistance (R_B) with the following relationship applying:

$$V_B = \frac{V_B}{I_B}$$

However, when the transistor is actually working the presence of an input excites an amplified current in the output, this current being equal to $I_B \times h_{fe}$. The effective emitter resistance is thus modified by feedback and becomes $R_B + R_E \times h_{fe}$. Thus the total effective input resistance is:

$$R_B + R_E (I + h_{fe})$$

The effect of R_E can be ignored as it is working as a reversed junction where its resistance is high enough not to have a shunting effect on the input circuit. This relationship only holds for as long as the output load is zero. Under realistic working conditions with a positive load the collector current (I_E) is *less* than $I_B \times h_{fe}$ and so the feedback is less. In other words the effective input resistance *decreases* with increasing loads. The amount of which this resistance is reduced is directly related to the I/R_c and the load inductance represented by I/R_L.

To simplify calculation both the feedback and conductance effects can be ignored, using the original equation:

$$\text{input resistance} = R_B = \frac{V_B}{I_B}$$

Both current gain and voltage gain are also affected by feedback, but again ignoring this for simple calculation:

$$\text{current gain} = h_{fe}$$

$$\text{voltage gain} = \frac{h_{fe} \times R_L}{R_B}$$

These formulas will exaggerate both current gain and voltage gain, but not to any great extent under normal working conditions. Certainly the difference will usually be within the spread of characteristics of individual transistors of the same type.

THE PRACTICAL CIRCUIT

A working circuit requires that the transistor be *biased* in a suitable manner to operate at its design point. This bias can be obtained from a separate battery, or more conveniently from the collector supply voltage. The simplest way this can be done is to tap the collector supply via a bias resistor (R_B), as in Fig. 10-3. The value of bias resistor required is normally:

$$R_B = \frac{V_{CC} - V_{BE}}{\text{bias required to give design base current } (I_B)}$$

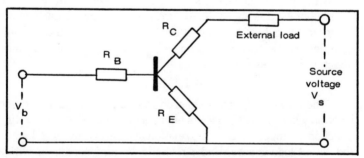

Fig. 10-2. Resistances effective in common emitter configuration.

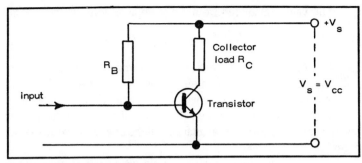

Fig. 10-3. Simple bias circuit.

Here V_s is the supply voltage. V_{BE} is the base-emitter voltage. In practical circuits this is usually of the same order as the 'starting' voltage of a transistor, e.g., 0.2 volts for a germanium transistor and 0.7 volts for a silicon transistor with a possible spread of 0.05 volts on either side.

Knowing the bias current required (from a transistor's characteristics), which will usually be in *microamps* for small transistors, and using numerical values for the voltages, the value of the bias resistor is given in *megohms* (M ohms). Not knowing the transistor characteristics, a typical value could be used, e.g., 30 μA for a small transistor. Calculation can be simplified further by ignoring V_{BE}. More accurately, the bias resistance should be calculated from the desired value of the emitter current (I_E):

$$I_E = \left(\frac{V_S - V_{BE}}{R_B} \right) (I + h_{fe})$$

$$\text{where, } R_B = \frac{(V_S - V_{BE})(I + h_{fe})}{I_E}$$

Again V_{BE} can be ignored to simplify calculation, when:

$$R_B = \frac{V_S \times h_{fe}}{I_E}$$

For example, using a silicon transistor with a nominal h_{fe} of 100, a supply voltage of 9 volts, and a desired emitter current of 1 mA:

$$R_B = \frac{9 \times 100}{0.001}$$
$$= 900 \text{ k ohm}$$

The value calculated from the 'complete' formula will be:

$$\frac{8.3 \times 101}{0.001} = 838 \text{ k ohm}$$

Nearest preferred value (i.e., actual resistance values obtainable) would be 820 k ohms in either case. In other words, the two different calculations both specify the same (practical) resistor value. The difference between the simplified formula calculation and the complete formula calculation will be even less in the case of a germanium transistor.

Operating parameters with this circuit are:

$$\text{base current } I_B = \frac{V_S - V_{BE}}{R_B}$$

Thus, emitter current is:

$$I_E = \frac{V_S - V_{BE}}{R1} \; (1 + h_{fe})$$

In this case the variations in working current (I_E) are substantially dependent on the spread of h_{fe}, thus this method of bias is really only suited to transistors which have a low h_{fe} spread.

The emitter current is also very much influenced by changes in supply voltage. However, this effect can be offset to a large extent by including a resistor of high value in the supply line, the higher the value used the less dependent the emitter current will be on the supply voltage. Biasing from a separate supply provides a more direct solution (Fig. 10-4). The value of bias resistor sequence (R_B) is given by:

$$R_B = \frac{V_{BB} - V_{BE}}{I_B}$$

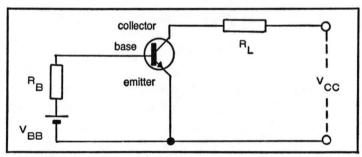

Fig. 10-4. Bias from a separate supply.

This formula can be simplified in the assumption that V_B is approximately 0.3 volts for a germanium transistor and 0.7 volts for a silicon transistor thus:

$$R_B \text{ for germanium transistor } = \frac{V_{BB} - 0.3}{I_B}$$

$$R_B \text{ for silicon transistor } = \frac{V_{BB} - 0.7}{I_B}$$

The only other requirement is that the polarity of the bias battery must be such as to provide the correct *direction* of the bias, i.e., collector-base reverse biased and base-emitter forward biased for an npn transistor; and collector-base forward biased at base-emitter reverse biased for a pnp transistor.

COMPLETE BIAS CIRCUIT

A more practical circuit which has a capability for stabilizing the operating point of the transistor is shown in Fig. 10-5. Here there are effectively two bias resistors, R1 and R2, working as a potential divider. In this case the bias voltage (V_B) developed is given by:

$$V_B = R_2 V_{cc} (R1 + R2)$$

There is also a voltage developed across R2 by the base current but this can be ignored for normal calculations. The voltage at the emitter is then given by:

$$V_E = V_B - V_{BE}$$
$$= \frac{R2\ V_{cc}}{(R1 + R2)} - V_{BE}$$

To accommodate spread in transistor characteristics V_E should be large compared with changes in V_{BE}. Also R_E must be large to 'stabilize' the emitter current against variations in the supply voltage (V_s). In practice a voltage drop (V_E) of about 1 volt should be allowed in the case of germanium transistors, and a voltage drop up to 3 volts allowed in the case of silicon transistors, to stabilize the emitter circuit effectively.

On the other hand, the values of R1 and R2 should not be so high that the base voltage is changed to a large extent by variations in base current. At the same time the values of R1 and R2 must be high enough not to waste power or drain power from the input signals.

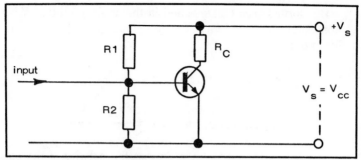

Fig. 10-5. Most commonly used bias circuit.

A possible amendment to this circuit is the inclusion of a resistor (R_E) in the emitter lead (together with a capacitor in parallel). The emitter current (bias current) is then given by:

$$I_E = \quad V_E/R_E$$
$$= \frac{V_{CC} \times R2}{R_E (R1 + R2)} \quad \frac{V_{BE}}{R_E}$$

In this type of circuit the combined parallel resistance of R2 and R_E should ideally be less than the value of R1. However, this is likely to result in a considerable loss of input signal, so a compromise solution is called for. A suitable starting point is to work R2 about four times the impedance of the transistor. The value of the parallel capacitor needs to be high (e.g., 100 μF or more for audio amplifiers). It is used to decouple R_E and prevent loss of gain due to negative feedback and in many cases may be omitted with no loss of performance.

With this form of bias the bias voltage (V_B) is determined by the values of resistors R1, R2 and the supply voltage (V_S), as long as the base current does not load the potential divider R1, R2:

$$V_B = \frac{R2 \times V_S}{R1 + R2}$$

The emitter voltage is given by:

$$V_E = \frac{R2 \times V_S}{R1 + R2} - V_{BE}$$

The emitter current of the transistor is thus:

$$I_E = \frac{R2 \times V_S}{R_E (R1 + R2)} - \frac{V_{BE}}{R_E}$$

In order to nullify the effect of spreads of transistor characteristics, and changes due to temperature effects, V_{BE} must be kept as constant as possible, which means that V_E should be large in comparison with V_{BE}. Equally, to nullify changes in supply voltage affecting I_E, R_E must also be large. In practice this demands a value of V_E of about 1 volt for germanium transistors, and a V_E of about 3 volts with silicon transistors. This desirable mode of working is also achieved if:

$$R_E \geq \frac{R1 \times R2}{(R1 + R1)(1 + h_{fe})}$$

This is the same as saying that for the emitter current to be independent of h_{fe}

$$V_E \geq \frac{R1 \times R2 \times I_B}{(R1 + R2)}$$

A great advantage of this type of circuit is that provided the above conditions are met the operating current I_E is independent of h_{fe}. Thus, it can adjust automatically to any spread of transistor characteristics and maintain the desired collector current.

It has been assumed that the voltage developed across the resistor R2 by the base current I_B is negligible, and, if this is true, it is seen that the operating current I_E is independent of h_{fe}, and therefore spreads in h_{fe}. If this voltage is not negligible, however, it can be shown that the operating current is given by:

$$I_2 = \frac{\dfrac{V_S R2}{R1 R2} - V_{BE}}{\dfrac{R1 + R2}{(R1 + R2)(1 + h_{fe})} + R_E}$$

and the current is then dependent on h_{fe}.

NOTES ON SILICON PLANAR TRANSISTORS

The base-emitter voltage of silicon transistors at low currents is 0.6 to 0.7 V, and of germanium transistors, 0.1 to 0.2 V. This is equivalent to a difference in 'starting' voltage, but the rate of change of current with base-emitter voltage above this 'starting' voltage is identical for silicon and germanium transistors.

The collector leakage current in silicon transistors is negligible up to the highest temperatures likely to be encountered in radio receivers. Consequently, it would appear possible to use a large dc base source resistance, and also to dispense with the normal stabilizing circuit, which applies negative dc feedback to the operating conditions. However, other effects need consideration and these may necessitate retaining the stabilizing circuit and limiting the magnitude of the source resistance. These effects are:

(i) Variations in supply voltage.

(ii) Spreads in the current amplification factor h_{fe}, and the base-emitter voltage V_{BE}.

(iii) Variations in V_{BE} and (to a lesser extent) h_{fe} with temperature.

Methods used (in order of preference) are:

(i) Conventional voltage biasing with emitter feedback (Fig. 10-5).

(ii) Current bias with collector feedback (Fig. 10-6).

(iii) Simple current biasing (Fig. 10-7).

Simple current biasing uses only one resistor (R1), when the base current is given by:

$$I_B = \frac{V_S - V_{BE}}{R1}$$

Thus the emitter current is given by:

$$\frac{V_S - V_{BE}}{R1} (1 + h_{fe})$$

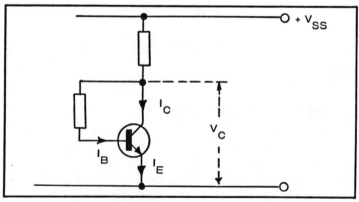

Fig. 10-6. Current bias with collector feedback.

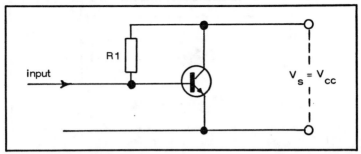

Fig. 10-7. Simple current biasing.

As an example if V_S is 7 V, and the desired emitter current is 1 mA, then for a transistor with a nominal value of h_{fe} of 100 (h_{fe} spread, 50 to 150), and a nominal value of V_{BE} of 0.7 V (V_{BE} spread, 0.65 to 0.75 V). R1 is 680 kΩ; the nominal value of V_{CE} is 7 V (= V_S); and the actual nominal I_E is 0.935 mA.

Substituting these values in the appropriate equation gives maximum and minimum values of I_E, resulting from the spread in h_{fe}, of 1.4 mA and 0.47 mA. Thus the spread in I_E resulting from the spread in h_{fe} is ± 49 percent.

The maximum and minimum values of I_E, resulting from the spread in V_{BE}, are 0.944 mA and 0.928 mA. Thus the spread in I_E resulting from the spread in V_{BE} is + 0.86 percent.

If V_S falls from 7 V to 3.5 V, the nominal emitter current falls from 0.935 mA to 0.41 mA; that is, I_E falls by 56 percent. However, because there are no resistors in either the collector or emitter leads, the collector-emitter voltage is equal to the supply voltage. Thus a higher-value series resistor can be used in the supply line (shared between all the rf and i-f stages), if the collector-emitter voltage in this method is to be the same as that in the other two biasing methods discussed; the higher the value of this series resistor the less dependent the emitter current will be on supply voltage.

The choice of a biasing arrangement depends very much on the supply voltage, the proportion of this voltage that can be dropped across a feedback resistor, and the type of transistor used.

Typical spreads in emitter operating current (I_E) resulting from spreads in h_{fe} and V_{BE} are:

Voltage biasing with emitter feedback:

$$h_{fe} \quad + 1 \text{ percent to } -1.6 \text{ percent}$$
$$V_{BE} \quad + 1.7 \text{ percent to } -1.7 \text{ percent}$$
$$V_{cc} \quad -62 \text{ percent}$$

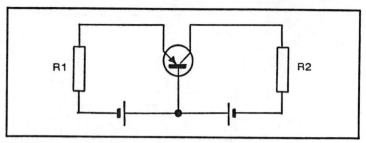

Fig. 10-8. Biasing with common base configuration.

Current biasing with collector feedback:

$$h_{fe} \quad +23 \text{ percent to } -35 \text{ percent}$$
$$V_{BE} \quad +1 \text{ percent to } -1 \text{ percent}$$
$$V_{cc} \quad -55 \text{ percent}$$

Simple current biasings:

$$h_{fe} \quad +50 \text{ percent to } -50 \text{ percent}$$
$$V_{BE} \quad +1 \text{ percent to } -1 \text{ percent}$$
$$V_{cc} \quad -50 \text{ percent}$$

BIAS FOR OTHER CONFIGURATIONS

Alternative transistor circuit configurations are *common-base* and *common-collector* (or emitter follower). They are normally only used in specialized amplifier circuits where their characteristics are advantageous—e.g., common collector made in cascaded amplifier stages because of their impedance characteristics and zero phase change from input to output.

Biasing with *common-base* configuration requires two separate voltage sources, one for output power and one for bias (Fig. 10-8). Input and output currents differ by the amount of bias current in the

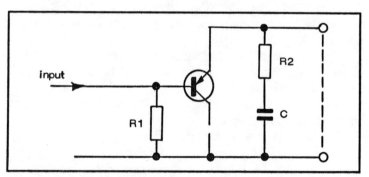

Fig. 10-9. Biasing with common collector configuration.

base circuit. Stabilization can be provided by feedback of collector circuit via a resistor (R2).

Biasing with *common-collector* configuration can be provided by a resistor (R1) across the input and a second resistor (R2) in series with a battery (or separate voltage source) across the output (Fig. 10-9). Resistor R2 is then, in effect, a load resistor common to both circuits. This method of biasing is very stable.

TRANSISTOR WORKING POINT

The basic diagram for a transistor working in common-emitter configuration is given in Fig. 10-10. Relating transistor characteristics to current design values will establish the working part of the transistor.

Transistor *input* characteristics are given by the relationship between base current (I_B) and base-emitter voltage (V_{BE}), with a characteristic wire for each value of collector voltage (V_{CC}). Base current is normally specified in microamps (μA), base-emitter voltage in millivolts (mV) and supply voltage (collector voltage) in volts (Fig. 10-11).

Transistor *output* characteristics are standardized by the variation in base current (I_B) with collector current (I_C) over a range of working collector-emitter voltages (V_{CE}). Collector current is normally specified in milliamps (mA), base currents in micro amps (μA) and collector-emitter voltage in volts (Fig. 10-12).

Load line equations can be derived from Fig. 10-10. On the *input* side we are concerned with I_B versus V_{BE}. Since a load line is linear, calculation of 'end' values enable the load line to be plotted.

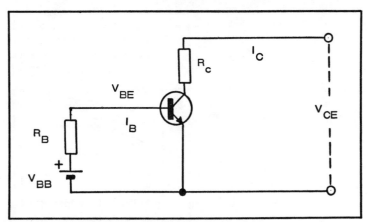

Fig. 10-10. Establishing the working point of a transistor.

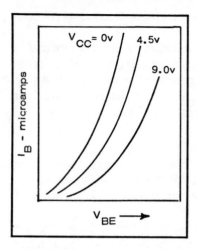

Fig. 10-11. Transistor input characteristics.

One 'end' value is obviously $I_B = 0$, when $V_{BE} = V_{BB}$. The other 'end' value is $V_{BE} = 0$, when $I_B = V_{BB}/R_B$.

Example: allocating the values $R_B = 10\,k$, and $V_{BB} = 2$ volt in Fig. 10-10 we have (on the input side):

$$\begin{aligned}
\text{at } I_B &= 0, \; V_{BE} = 2 \text{ volts} \\
\text{at } V_{BE} &= 0, \; I_B = 1/10000 \\
&= 100 \; \mu A
\end{aligned}$$

These points can be plotted on an appropriate transistor characteristic input graph and the load line formed by connecting the two points with a straight line. The operating point is then established by the point at which the load line crosses the appropriate

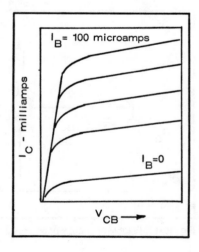

Fig. 10-12. Transistor output characteristics.

transistor curve (i.e., V_{CC} curve) as in Fig. 10-13. This gives actual working values for I_B and V_{BE}.

On the *output* side we are concerned with I_C *versus* V_{CE}. The basic relationship (from Fig. 10-10) is:

$$V_{CC} = I_C R_L + V_{CE}$$

Again taking 'end' values:

$$\text{when } V_{CE} = 0, \ I_C = V_{CC}/R_L$$
$$\text{when } I_C = 0, \ V_{CE} = V_{CC}$$

End points for the load curve are then calculated as before, taking appropriate values.

Example: allocating the values $V_{CC} = 10$ volts and $R_L = 3k$

$$\text{at } V_{CE} = 0, \ I_C = 10/3{,}000$$
$$= 3.3 \text{ milliamps}$$
$$\text{at } I_C = 0 \quad V_{CE} = 10 \text{ volts}$$

This plot is shown in Fig. 10-14, superimposed on a typical transistor characteristic graph. Again the working point of the transistor is established by the intersection of the load line with the characteristic curve, giving working values of I_C and V_{CE}.

At the same time the emitter current can be calculated. This is given by:

$$I_E = I_B + I_C$$

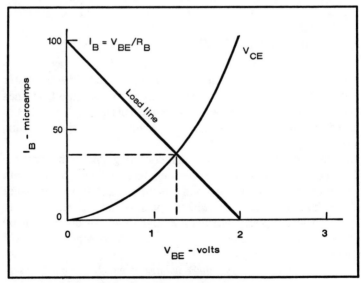

Fig. 10-13. Establishing the transistor operating point.

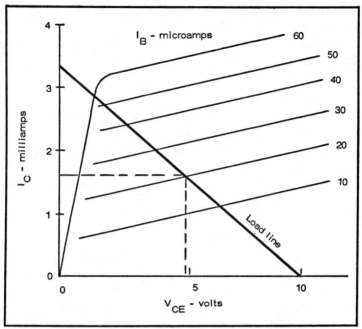

Fig. 10-14. Load line plotted from $V_{CE} = 0$ to $I_C = 0$.

Both I_B and I_C have been determined by superimposition of the load curves on the input and output characteristic curves, respectively.

TRANSISTOR CIRCUIT DESIGN

Transistor circuit *design* then starts the other way round—selecting a suitable or required, working point on the transistor characteristic curves and adjusting values to make the load lines pass through the design operating points. This is simply a matter of substitution in the appropriate formula(s) for transistor biasing circuits.

In practice, selecting a design operating point is not particularly critical. Also the actual operating point will tend to shift with temperature. The only real limits to positioning of the working point are within the *maximum ratings* specified for the transistor in question—particularly *maximum reverse voltage* or bias (V_{EB} max) *maximum collector current* (I_C max), and maximum total power dissipation (P_t max).

There is also an *avalanche voltage* at which a transistor will break down or 'run away' and destroy itself. This can be a particularly significant factor in pulsed circuits. Highest maximum collector voltage ratings for a transistor normally relate to maximum collection-emitter voltage at a specified base resistance. Lowest maximum collector voltage ratings are normally for reverse (bias) voltage, which will be a substantially lower figure. A general value for a 'safe' working voltage is a figure about mid-way between the two.

STABILIZING THE WORKING POINT

The actual working point or Q-point of a transistor may shift up or down the load line due to parameter charges (or difference in parameters when one transistor is changed for another of the same type). Specifically, for instance, an increase in temperature will normally produce an increase in collector current (I_C), with a corresponding reduction in collector voltage (V_C); and vice versa. The actual working part may therefore range up and down the load line (Fig. 10-15).

The circuit designer normally tries to accommodate this in the biasing circuit, the aim being to change the base current automatically so that the working point remains fixed regardless of changes in I_C, h_{fe}, V_{BE} or V_{CC}.

At this point it is appropriate to summarize the advantages and disadvantages of the alternative biasing methods previously described.

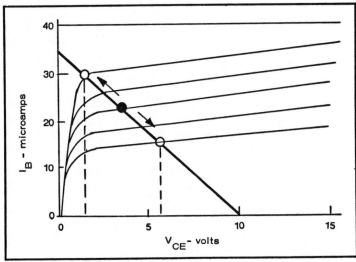

Fig. 10-15. Working point shift in non-stabilized circuit.

Current biasing—has the advantage of being simple and calling for a minimum number of components. Bias is largely independent of the base-emitter voltage V_{BE}, and if V_{BE} is made very much less than V_{CC} the bias current (I_B) is largely independent of V_{BE}. This will provide better temperature stabilization.

The disadvantages are:

(i) The current gain increases with temperatures and can vary widely with different transistors of the same or similar type. This can result in widely different working points than those predicted. To a large extent this effect can be reduced if resistive leads are kept small, and collector current made large.

(ii) The collector current is not controlled with reverse swing (i.e., when the collector is biased in the reverse direction. This can limit the swing, or even cause the transistor to 'bottom' at higher temperatures.

(ii) Thermal runaway can occur with simple collector feedback biasing if the collector voltage reaches too high a value.

Voltage biasing—can provide better stabilization (virtually all the stability necessary in straightforward amplifier circuits) at the expense of:

(i) Drawing a higher base current resistor and thus higher drain from the power supply.

(ii) Reduction of input resistance.

Temperature stabilization is, however, very good.

Chapter 11

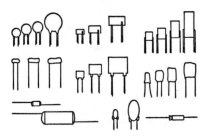

Field Effect Transistors

The *Field Effect transistor* (FET) is a different type of semiconductor transistor with characteristics more like a thermionic valve than a bipolar transistor. Its correct definition is a *unipolar* transistor. The way in which it works can be understood by presenting it in electronic 'picture' form as in Fig. 11-1 where it can be seen that it consists of a *channel* of either p-type or n-type semiconductor material with a collar or *gate* of opposite type material at the center. This forms a semiconductor junction at this point. One end of the channel is called the *source* and the other end the *drain*.

A FET is connected in a similar manner to a bipolar transistor, with a *bias* voltage applied between gate and source, and a supply voltage applied across the center of the channel (i.e., between source and drain). The source is thus the common connection between the two circuits. Compared with a bipolar transistor, however, the bias voltage is *reversed*. That is, the n-gate material of a p-channel FET is biased with *positive* voltage; and the p-gate material of an n-channel FET is biased with negative voltage (Fig. 11-2). This puts the two system voltages 'in opposition' at the source, which is responsible for the characteristically *high input resistance* of FETs.

The effect of this reverse bias is to form an enlarged depletion layer in the middle of the channel, producing a 'pinching' effect on the flow of electrons through the channel and consequently on the current flow in the source-to-drain circuit. If enough bias voltage is applied, the depletion layer fills the whole gate ('shuts the gate'),

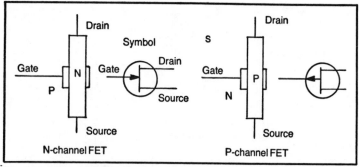

Fig. 11-1. Field effect transistors doping.

causing *pinch off*, when the source-to-drain current falls to zero (in practice nearly to zero, for there will still be some leakage). With *no* bias applied to the gate, the gate is 'wide open' and so maximum current flows.

In effect, then, the amount of reverse bias applied to the *gate* governs how much of the gate is effectively open for current flow. A relatively small change in gate voltage can produce a large change in source-to-drain current, and so the device works as an amplifier. In this respect a p-channel FET works very much like a pnp transistor, and an n-channel FET as an npn transistor. Its main advantage is that it can be made just as compact in size, but can carry much more power. In this respect—and the fact that it has a high input resistance, whereas a bipolar transistor has a low input resistance—it is more like a thermionic value in characteristics than a bipolar transistor. It also has other advantages over a bipolar transistor, notably a much lower inherent 'noise' making it a more favorable choice for an amplifier in a high quality radio circuit.

The type of field effect transistor described is correctly called a junction field effect transistor, or JFET. There are other types

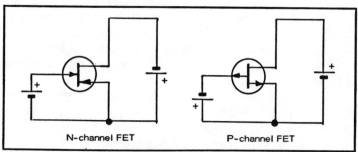

Fig. 11-2. N-channel and p-channel FETs (diagrammatic).

produced by modifying the construction. The insulated gate field effect transistor of IGFET is self-explanatory. The IGFET has even higher input resistance (because the gate is insulated from the channel), and is also more flexible in application since either 're-verse' or 'forward' polarity can be applied to the gate for bias. FETs, of either type, can also be made with two gates. In this case the first gate becomes the signal gate (to which the input signal is applied) and the second gate becomes the *control gate*, with similar working to a pentode value.

FETs are also classified by the mode in which they work. A JFET works in the *depletion mode*, i.e., control of the extent of the depletion layer, and thus the 'gate' opening being by the application of a bias voltage to the gate. An IGFET can work in this mode, or with opposite bias polarity, in which case the effect is to produce an increasing 'gate opening', with enhanced (increased) source-to-drain current. This is called the enhancement *mode*. An FET de-signed specifically to work in the enhancement mode has no channel to start with, only a gate. Application of a gate voltage causes a channel to be formed.

The basic circuit of an FET amplifier is very simple (Fig. 11-3 with polarity drawn for a p-channel FET). Instead of applying a definite negative basis to the gate, a high value resistor R1 is used to maintain the gate at substantially zero voltage. The value of resistor R2 is then selected to adjust the potential of the source to the required amount *positive* to the gate. The effect is then the same as if negative bias were applied direct to the gate. This arrangement will also be self-compensating with variations in source-to-drain cur-rent. The third resistor R3 is a load resistor for the FET to set the design operating current. Capacitor C1 acts as a conductive path to remove signal currents from the source.

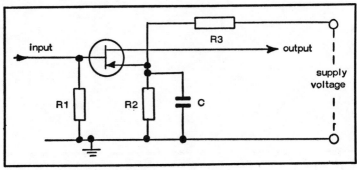

Fig. 11-3. Basic FET amplifier.

Both junction-type (JFET) and insulated gate (IGFET) field effect transistors are widely used, the latter having the wider application, particularly in integrated circuits. The metal-oxide semiconductor FET, generally referred to as a MOSFET, can be designed to work in either mode, i.e., as a depletion MOSFET, or enhancement MOSFET. The former is usually an n-channel device and the latter a p-channel device. P-channel MOSFETs working in the enhancement mode are by far the more popular, mainly because they are easy to produce. In fact, an n-channel MOSFET can be made smaller for the same duty and has faster switching capabilities, and really is to be preferred for LSI MOS systems.

FET CONFIGURATIONS

Alternative configurations for FETs used in low frequency circuits are common source (CS) or common drain (CD) (Fig. 11-4). In the common source configuration the output is taken from the drain when the source resistance is effectively zero. With common drain configuration the output is taken from the source and the drain resistance is effectively zero. The sequel source resistance is not important in either configuration since it is in series with the gate, which draws negligible current. In practical circuits suitable biasing must be incorporated to fix the operating and provide linear operation.

BIASING FETS

Simple (source) self-biasing can be used with FETs or depletion-mode MOS devices, as shown in Fig. 11-5. This circuit

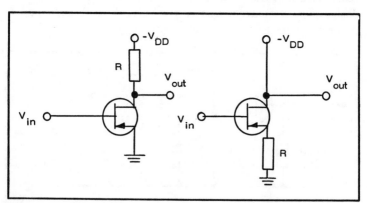

Fig. 11-4. Common source (CS), left; and common drain (CD), right configurations.

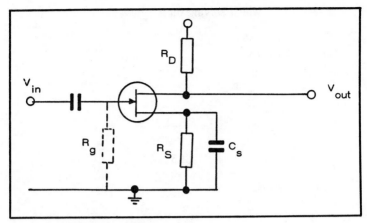

Fig. 11-5. Simple current biasing.

also has the advantage that necessary component values can be calculated quite easily. Basic requirements are to bias the circuit at a specified current T_{DS} with a given V_{DD}, biasing V_p and I_{Dss} from the transistor characteristics.

Design equations are:

(i)
$$I_{DS} = I_{DSS} \left(1 - \frac{V_{GS}}{V_p}\right)^2$$

(ii)
$$g_{mo} = \frac{-2I_{DSS}}{v_p}$$

(iii)
$$g_m = g_{mo} \left(1 - \frac{V_{GS}}{V_p}\right)$$

(iv)
$$A_V = g_m R_D$$

Example: In the basic amplifier circuit of Fig. 11-5 where $V_{DD} = 24$ V and it is desired to bias the circuit at $I_{DS} = 0.8$ mA. A voltage gain of at least 20 dB is required. For the transistor chosen $V_p = 2.0$ V and $1_{DSS} = 1.65$ mA.

Using equation (i):

$$I_{DS} = I_{DSS} \left(1 - \frac{V_{GS}}{V_p}\right)^2$$

and substituting

$$0.8 = 1.65 \ (1 + V_{GS}/2)^2$$

which solving for V_{GS} gives

$$V_{GS} = -0.62 \text{ V}$$

Using equation (ii):

$$g_{mo} = \frac{-2I_{DSS}}{v_p}$$

$$= \frac{2 \times 1.65}{2}$$

$$= 1.65 \text{ mA/V}$$

Using equation (iii):

$$g_m = g_{mo} \quad (1 - \frac{V_{GS}}{V_p}$$

$$= 1.65 \quad (1 - \frac{0.62}{2.0}$$

$$= 1.14 \text{ mA/V}$$

$$\text{Now } R_S = \frac{-V_{GS}}{I_D} = \frac{0.62}{0.8}$$

$$= 770 \text{ ohms}$$

R_S then needs to be bypassed by a large capacity say 10 μF.

Using equation (iv): A voltage gain of 20 dB corresponds to a voltage gain of 10. Hence from equation (iv) we want:

$$g_m R_d \quad 10$$

$$\text{or} \quad E_D \ 10/gm$$

$$10/.00114$$

$$8.76 \text{ k ohms}$$

FIXED-BIAS GATE

Sometimes it may be desirable, or even necessary, to provide a fixed bias on the gate as well as self-bias. This has the useful effect of compensating for variations in the FET characteristics. This can be done via a separate supply voltage in series with R_G, but more conveniently from just one voltage source as in Fig. 11-6. Design formulas to use in this case are:

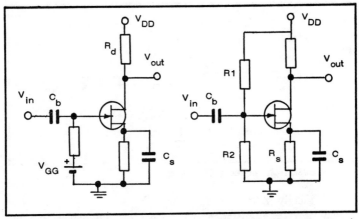

Fig. 11-6. Fixed bias plus self bias (left); and bias from single power supply (right).

$$V_{GG} = \frac{R_2 V_{DD}}{R1 + R2}$$

$$R_G = \frac{R1R2}{R1 + R2}$$

HIGH FREQUENCY CS AMPLIFIER

The basic configuration for a common source (CS) amplifier at high frequencies is shown in Fig. 11-7 where the output voltage is equal to I_z where:

I is the short circuit current

Z is the impedance seen between the output terminals

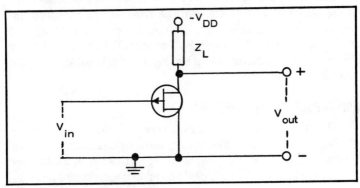

Fig. 11-7. Basic high frequency CS amplifier.

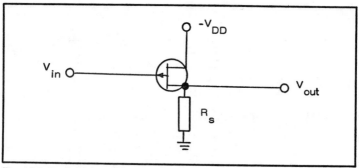

Fig. 11-8. Basic high frequency CD amplifier.

Specifically, $Z = Y_L + Y_{DS} + g_d + Y_{gd}$

where Y_L = admittance (reciprocal of impedance) corresponding to Z_L

Y_{DS} = admittance corresponding to C_{DS}

g_d = conductance corresponding to r_d (drain resistance)

V_{gd} = admittance corresponding to C_{GD}

This takes into account the capacitance effects present between source and drain and gate and drain which are ignored for low frequency working. The gate-drain capacitance C_{GD}, in fact, effectively connects the gate and drain and for an FET to have negligible input admittance over a wide range of frequence, but the gate-drain capacitance C_{GD} and the drain-source capacitance C_{DS} must be negligible.

Input capacitance is particularly significant in the case of cascaded amplifiers where the input impedance of the second stage acts as a shunt across the first output stage, and so on. Since the reactance of a capacitor decreases with increasing frequency, output impedance of the first stage will be lowered for higher frequencies, resulting in a decreasing gain at higher frequencies. This circuit can be worked by either JFET or MOSFET devices, the only difference being in the method of biasing. Biasing is not shown in the basic circuit.

HIGH FREQUENCY CD AMPLIFIER

The principle advantage of a common drain (CD) configuration for a high frequency amplifier is that the input capacitance is lower than for a CS configuration (Fig. 11-8). Again this circuit can be used with both JFET or MOSFET devices, adding appropriate biasing.

Chapter 12

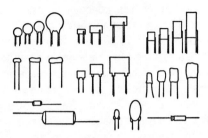

Integrated Circuits

The basic element of an integrated circuit or IC is a single *chip* of silicon about 0.050 inches square which can contain upwards of 50 separate components, interconnected to form complete circuits. In manufacture up to 400 chips may be formed as a *wafer* only 1 inch square, and up to 10 wafers processed at the same time. Four thousand ICs produced simultaneously, containing the equivalent of nearly a quarter of a million components—and that is only *small-scale integration* (SSI).

It is possible to achieve much higher component densities. With medium-scale integration (MSI), component density is greater than 100 components per chip; and with large-scale integration (LSI), component density may be as high as 1000 or more components per chip. Both MSI and LSI are extensions of the original integrated circuit techniques using similar manufacturing methods. The only difference is in the matter of size and physical separation of the individual components and the method of interconnection.

IC COMPONENTS

Transistors and *diodes* are formed directly on the surface of the chip with their size and geometry governing their electrical characteristics as well as density level, etc. Where a number of such components are involved in a complete integrated circuit, their performance is usually better than that of a circuit with discrete (separate) components because they are located close together and their electrical characteristics are closely matched.

Resistors can be formed by silicon resistance stripes etched in the slice, or by using the bulk resistivity of one of the diffused areas. There are limits, however, to both the range and tolerance of resistance values which can be produced by these methods. 'Stripe' resistors are limited to a minimum width of about 0.025 mm (0.001 in) to achieve a tolerance of 10 percent. Practical values obtained from diffused resistors range from about 10 ohms to 30 k ohms, depending on the method of diffusion with tolerance of plus or minus 10 percent. Better performance can be achieved with thin-film resistors with resistance values ranging from 20 ohms to 50 k ohms.

A method of getting round this problem when a high resistance is required is to use a transistor biased almost to cut-off instead of a resistor in an integrated circuit where a resistance value of more than 50 k ohms is required. This is quite economic in the case of integrated circuit manufacture and a method widely used in practice.

Capacitors present more of a problem. Small values of capacitance can be produced by suitable geometric spacing between circuit elements and utilizing the stray capacitance generated between these elements. Where rather higher capacitance values are required, individual capacitors may be formed by a reversed-bias pn junction; or as thin-film 'plate' type using a tiny aluminum plate and a MOS (metal-oxide-semiconductor) second plate. The former method produces a *polarized* capacitor and the thin film method a *non-polarized* capacitor. The main limitation in either case is the relatively low limit to size and capacitance values which can be achieved—typically 0.2 pF per 0.025 mm (0.001 inch) square for a junction capacitor and up to twice this figure with a thin film MOS capacitor, both with fairly wide tolerances (plus or minus 20 percent). Where anything more than moderate capacitor values are needed in an integrated circuit it is usually the practice to omit the capacitor from the circuit and connect a suitable discrete component externally.

Both resistors and capacitors fabricated in ICs also suffer from high temperature coefficients (i.e., working values varying with temperature) and may also be sensitive to voltage variations in the circuit.

Unlike printed circuits, it is not possible to fabricate inductors or transformers in integrated circuits at the present state-of-the-art. As far as possible, therefore, ICs are designed without the need for such components; or where this is not possible, a separate conventional component is connected externally to the integrated circuit.

From the above it will be appreciated that integrated circuits are quite commonly used as 'building blocks'" in a complete circuit, connected to other conventional components. A simple example is shown in Fig. 12-1 using a ZN414 as a basic "building block' in the construction of a miniature AM radio. Although a high gain device (typical power gain 72 dB) the integrated circuit needs a following stage of transistor amplification to power a crystal earphone; high value decoupling capacitors; and a standard coil and tuning capacitor for the tuned circuit. The complete circuit is capable of providing an output of 500 millivolts to the earphone, with a supply voltage of 1.3 V and a typical current drain of 0.3 milliamps.

IC PACKAGES

Many smaller ICs are 'packaged' in transistor shape 'cans' with leads emerging from the bottom. The number of leads in this case is usually three (confusing with a transistor!), six or eight. Leads are identified by numbers, reading around the bottom (normally in a clockwise direction). Other ICs come in the form of flat packages with leads emerging from each side. These are three different arrangements used (see also Fig. 12-2):

(i) *Dual-in-line*, where the leads on each side are bent down to form two separate rows to plug directly into a printed circuit panel or IC holder.

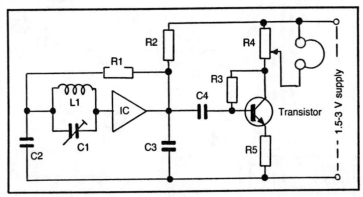

Fig. 12-1. AM radio circuit built around a single IC (ZN414). Component values: R1 – 100 k ohms; R2 – 1 k ohm; R3 – 100 k ohms; R4 – 10 k ohms; R5 – 100 k ohms; C1 – tuning capacitor; L1 matching coil; C2 .01 μF; C3 – 0.1 μF; C4 – 0.1 μF; transistor ZTX 300, phones, crystal earpiece.

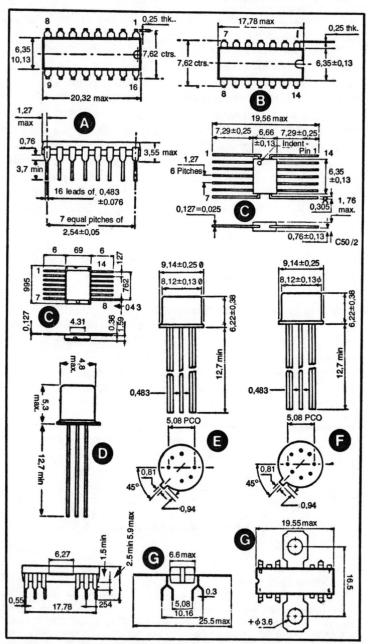

Fig. 12-2. Examples of integrated circuit outlines; A. 16-pin dual in-line, B. 14-pin dual in-line, C. flat (ceramic) package, D. 3-lead can, E. 6-lead can, F. 8-lead can, G. 12-pin quad in-line (with heatsink tabs).

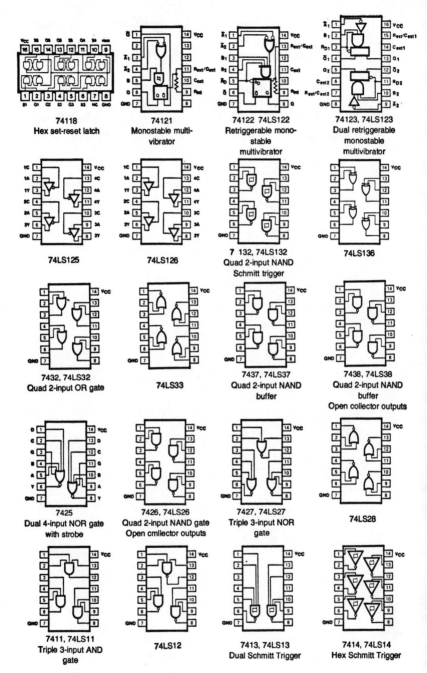

Fig. 12-3. Examples of ICs with pin numbering.

123

(ii) *Quad in-line*, like dual in-line, except that the leads on each side form two parallel rows.

(iii) *Flat*, where the leads emerge straight and from each side of the package.

In all cases leading numbering normally runs around the package, starting from top left and ending at top right. The number of leads may be anything from eight to sixteen or even more. Typical IC package forms, with lead numbering are shown in Fig. 12-3.

Chapter 13

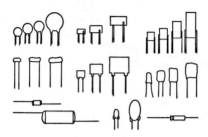

Heatsinks for ICs

For dissipating heat from ICs handling power up to about 3 watts, areas etched on the copper of a printed circuit board can conveniently be used as heatsinks. ICs which are suitable for heatsinks of this type are usually fitted with a tab or tabs for soldering directly to the copper lands forming the heatsink. To find the required copper area for a heatsink, first calculate the maximum power to be dissipated:

$$\text{Power (watts)} = 0.4 \frac{Vs^2}{8R_L} + V_s \cdot I_d$$

where V_s is the maximum supply voltage

I_d is the quiescent drain current in amps under the most adverse conditions

R_L is the load resistance (e.g., the loudspeaker resistance in the case of an audio amplifier circuit).

Strictly speaking the value of V_s used should be the battery voltage plus an additional 10 percent, e.g., if the circuit is powered by a 12-volt battery, the value of V_s to use in the formula is $12 + 1.2 = 13.2$ volts. This allows for possible fluctuations in power level, such as when using a new battery. If the circuit has a stabilized power supply, then V_s can be taken as this supply voltage.

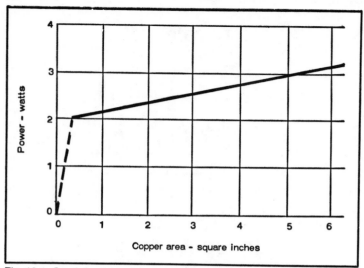

Fig. 13-1. Graph for determining copper area required for heatsinks.

The quiescent drain current (I_d) is found from the IC parameters as specified by the manufacturers and will be dependent on supply voltage. Figures may be quoted for 'typical' and 'maximum'. In this case, use the maximum values. Figure 13-1 can then be used to determine a suitable area of copper for the heatsink. This has been designed to provide enough copper area to limit the maximum temperature of the IC to 55° C, which is a safe limit for most ICs.

Example: Supply voltage for a particular IC is 12 volts. Load resistance is 4 ohms and the maximum quiescent current drain quoted for the IC at this operating voltage is 20 milliamps. The supply voltage is not stabilized, so the value to use for V_s is:

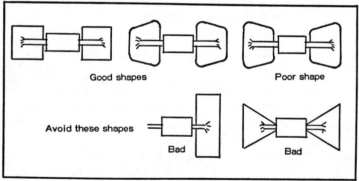

Fig. 13-2. Good and bad copper shapes.

$$12 + 1.2 = 13.2 \text{ volts}$$

$$\text{Thus power} = 0.4 \times \frac{13.2^2}{8 \times 4} + (13.2 \times 0.020)$$
$$= 2.178 + 0.264$$
$$= 2.422 \text{ watts (say 2.5 watts)}$$

From the graph a suitable copper area is seen to be approximately 1.6″ square (40 mm square). Suitable copper area shapes are shown in Fig. 13-2.

For dissipating powers above 3 watts it is more effective (and more convenient) to use an actual heatsink fitted to the IC itself. Matching heatsinks are readily available for ICs designed to handle high powers.

Chapter 14

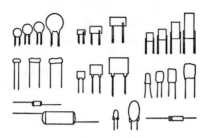

LEDs and LCDs

LEDs (light-emitting diodes) are simple and easy to work with. Specific values required relating to the LED are:

(i) Rated forward voltage at which the LED lights (typically 1.5 to 2 V).

(ii) Rated forward current of the LED at specified forward voltage current (typically 10 - 50mA, depending on size and types 20 mA is an average figure).

(iii) Maximum forward current specified (typically 50 mA or 100 mA).

(iv) Maximum reverse voltage, if known. Otherwise take as 2 volts.

For working in a *dc* circuit a ballast resistor is required to drop the voltage to the rated forward voltage (Fig. 14-1).

Formula for calculating value of ballast resistor:

$$\text{ballast resistor} = \frac{V_s - V_f}{I_f}$$

where V_s = *dc* supply voltage
V_f = rated forward voltage of LED
I_f = rated forward current of LED
at specified forward voltage

Example: An LED has a rated forward voltage of 2 V and rated forward current of 20 mA. Find the ballast resistor required when a 6 volt *dc* supply is present.

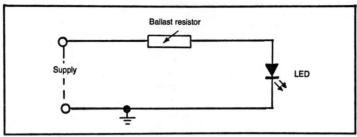

Fig. 14-1. Basic LED circuit with ballast resistor.

$$\text{ballast resistor} = \frac{6 - 2}{.020}$$
$$= 200 \text{ ohms}$$

Adjust to nearest preferred value, i.e., 180 ohms or 220 ohms. To safeguard this circuit against reverse voltage a diode can be incorporated either in series or back-to-back configuration (Fig. 14-2).

With an *ac* supply a diode is connected in back-to-back configuration. The value of the ballast resistor is calculated in the same manner as for the *dc* current, but need be only *one half* of this value.

Example: the same LED is to be worked from a 20 volt *ac* supply, using a diode in back-to-back configuration. Calculate the ballast resistor required:

$$\text{Ballast resistor} = \frac{1}{2} \times \frac{20 - 2}{.020}$$
$$= 450 \text{ ohms}$$

LEDs are very familiar in the form of groups, or LED displays, e.g., in calculators, digital instruments, etc. The most common form is a seven-segment display and associated decimal point. Such a

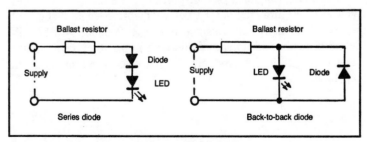

Fig. 14-2. Diode protection against reverse voltage.

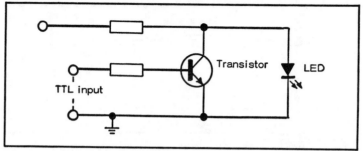

Fig. 14-3. LED display driven from a TTL device.

display can light up numerals from 0 to 9, depending on the individual segments energized, with or without the decimal point lighted. Each segment (or point) is, of course, an individual LED.

Specific advantages of LEDs are that they require only low voltages, are fast switching and can be produced in very small sizes, if required. The most widely used seven-segment displays, for example, give figures which are 0.3 to 0.5 inches high. Power consumption is relatively low, but an 8-digit seven-segment display could have a maximum power consumption of 8 × 7 × .020 amps at 2 volts or 2.24 watts. LED displays are commonly driven from a TTL device through a transistor buffer stage (Fig. 14-3).

Liquid crystal displays have the advantage that they need very much lower voltages and currents. Current drain can be as low as 1 μA per segment, or with a field effect liquid crystal as low as 0.2 μA. They have the further advantage that they can be driven directly by a MOS logic device without the need for an intermediate buffer/driver stage.

Bipolar liquid crystal displays commonly operate on a voltage of 25 V, draining 1 μA per segment. Power drain by a single 8-segment display (single numeral plus dot) would therefore be 8 × 25 × .000001 = .2 milliwatts

Field effect liquid crystal displays operate on lower voltages (typically 7 V), draining 0.3 μA per segment. A single 8-segment display would therefore consume 8 × 7 × .0000003 = .017 milliwatts. A further advantage of this type is that they give a black image on a light background, offering better contrast for seeing in poor light.

Chapter 15

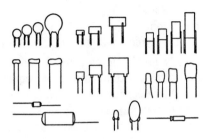

Crystals

Crystals of certain substances, notably *quartz*, possess what is known as *piezoelectric* properties. This means that they are capable of vibrating mechanically up to say high frequencies and at the same time generating an alternating voltage at vibration frequency between two opposite forces. Concisely, if an alternating voltage is applied to opposite faces of the crystal, it will vibrate at the ac frequency.

In hardware terms, crystals are cut to a specific geometry (determining their frequency), mounted between two metal plates and encased in hermetically sealed rectangular cans with two wires emerging from the base. They are most commonly cut for parallel resonance, but some (specifically those above 40 MHz) may be series resonance. A parallel resonance crystal can be used in a series resonant circuit simply by connecting a small (e.g., 65 pF) trimmer capacitor in series with the crystal.

Crystals are specified by type:

(i) Frequency standards—commonly with frequencies of 100 kHz, 1 MHz, and 10 MHz.

(ii) Microprocessor or crystals for use with the most popular microprocessor chips and in digital circuitry. Commonly these cover frequencies from 1 MHz to 18.432 MHz.

(iii) Color TV crystals for use in TV receivers, TV games, etc.

(iv) Radio control crystals—matched pairs (differing by the standard i-f) covering spot frequencies available in the 27 MHz, 35 MHz, 45 MHz, 70 MHz bands.

Crystal performance is usually specified by:

(a) Frequency
(b) Adjustment tolerance
(c) Temperature stability
(d) Temperature range
(e) Load capacitor (typically 30 pF or 32 pF for most types)

Although basically employed as a fixed frequency oscillator crystals may be subject to a certain amount of frequency drift with time (also if subjected to mechanical or thermal shock). Ultra-high frequency crystals may also require aging before reaching a stable condition—i.e., their frequency may change appreciably from the figure as originally manufactured over the first few months. (Aging can be accelerated by artificial means to reduce the effect of this limitation.)) Most crystal circuits are associated with a variable capacitor across the crystal for trimming but the adjustment of frequency possible is within very narrow limits—usually of the order of a few parts in 10,000.

Chapter 16

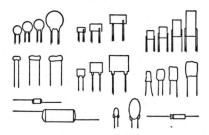

Op Amps

The integrated circuit operational amplifier or *op amp* is one of the most useful 'building blocks' available for electronic circuits. A typical op amp is, in fact, an amplifier circuit which can be made to perform a variety of linear functions.

A basic op amp is a three-terminal device with two inputs and one output. The input terminals are referred to as *inverting* and *noninverting*. More complex op amps have additional internal circuiting and further pins—for positive and negative supply, offset switch, ground, etc. They may, therefore, show up to fourteen leads (sometimes more), the designation of which is specific to that particular IC. Lead designation is not the same with different types and makes, although pins 2 and 3 are commonly used for inverting and noninverting inputs, respectively and 1 (or 6) for output.

BASIC OP AMP

The basic op amp (three terminal device) can work in an *inverting* or *noninverting* mode; or as a differential amplifier. In practical circuits it is associated with a feedback resistor R_F, and resistors on the input side. The basic circuit for operation in the inverting mode (input to inverting or − pin) is shown in Fig. 16-1, where:

$$V_{out} = -V_{in} \times R_F/R1$$

Note the inverted sign of the output.

The basic circuit for operating in the noninverting mode (input to noninverting or + pin) is shown in Fig. 16-2, where:

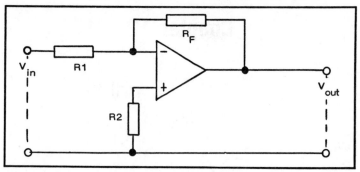

Fig. 16-1. Basic op-amp circuit, inverting mode.

$$V_{out} = V_{in} \times \frac{R2 + R_F}{R1}$$

The *differential amplifier* circuit of Fig. 16-3 has two inputs V1 and V2. In this case

$$V_{out} = V2 \left(\frac{1 + R3/R1}{1 + R2/R4} \right) - V2 \quad \frac{R3}{R1}$$

It is convenient to make R3/R1 equal to R4/R2. This working formula simplifies to:

$$V_{out} = \frac{R3}{R1} \ (V1 - V2)$$

This circuit gives an output depending on the difference between V1 and V2. If the input voltages are identical (V1 = V2), there is no output.

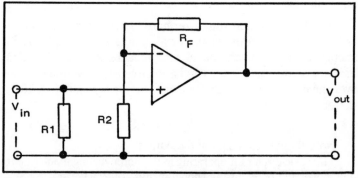

Fig. 16-2. Basic op-amp circuit, noninverting mode.

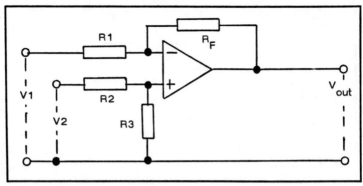

Fig. 16-3. Differential amplifier circuit.

ADDER CIRCUITS

A basic *adder* circuit is shown in Fig. 16-4—technically called a summing amplifier. The performance is given by:

$$V_{out} = -\left(\frac{RF}{R1} \times V1 + \frac{RF}{R2} \times V2 + \frac{RF}{R3} \times V3 + ... \frac{RF}{R_N} \times V_N\right)$$

or more simply:

$$V_{out} = -RF\left(\frac{V1}{R1} + \frac{V2}{R2} + \frac{V3}{R3} + ... \frac{V_N}{R_N}\right)$$

If all the resistors are given the same value (R1 = R2 = R3 etc.):

$$V_{out} = -\frac{RF}{R1}(V1 + V2 + V3 \text{ etc, i.e., directly 'adds' V1, V2, V3, etc.}$$

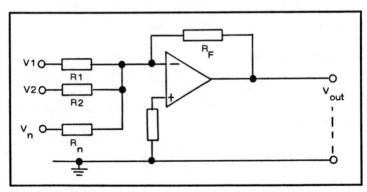

Fig. 16-4. Basic adder circuit.

135

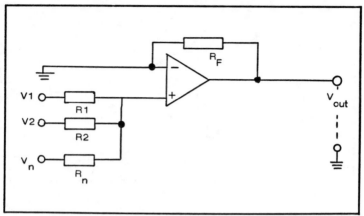

Fig. 16-5. Noninverting adder.

Better still, make RF = R1 so that RF/R1 = 1, when the output is equal to the sum of the input voltages:

$$V_{out} = V1 + V2 + V3 + \ldots V_n$$

Equally, by using different values for R1, R2, etc. multiplies or fractions of the individual input signal voltages can be summed. For example, to sum V1 + ½ V2 + ¼ V3, resistor R2 would be determined to drop V2 to ½ V2. Similarly resistor R3 would be determined to drop V3 to ¼ V3.

This circuit is very useful for it can accommodate a large number of inputs for summing, each requiring the addition of one resistor. Also, since the (+) input is grounded there is a minimum of interaction between the input sources. Note, however, that the sign of the output is reversed.

By connecting in the opposite mode (Fig. 16-5) the circuit will work as a noninverting adder where input and output have the same polarity of signal and are thus in phase.

ADDER/SUBTRACTOR

A combination of noninverting and inverting mode circuitry can be used to produce an *adder-subtractor* (Fig. 16-6). If all the resistors are made the same value, i.e., R1 = R2 = R3 = R4, we have on the inverting side:

$$V_{out} = -\frac{R_F}{R1} (V1 + V2)$$

and on the non-inverting side:

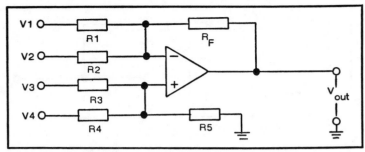

Fig. 16-6. Adder-subtractor circuit.

$$V_{out} = \frac{R_F}{R3} \ (V3 + V4)$$

The complete output is thus:

$$V_{out} = \frac{R_F}{R1} \ (V3 + V4 - V1 - V2)$$

Additional inputs can be added on either the inverting or noninverting side. All inputs on the noninverting side are summed in passing to the output—all inputs on the inverting side subtract from the output voltage.

The value of R1, etc., is chosen to match the op-amp characteristics. R5 can be given the same value as R3. The value of the feedback resistor R_F should be the same as R1.

MULTIPLIER

A multiplier circuit is shown in Fig. 16-7. The gain of this circuit is given by $-R_F/R1$. Thus if R_F is twice the value of R1, the

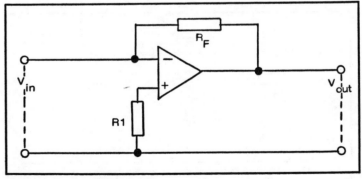

Fig. 16-7. Basic multiplier circuit.

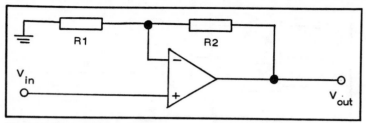

Fig. 16-8. Noninverting multiplier.

input signal is inverted and multiplied by 2. For any other multiplication factor, the value of $R_F/R1$ is adjusted accordingly, bearing in mind that resistors are only available in preferred numbers.

The accuracy of such a multiplier depends on the accuracy of the resistance values, calling for the use of precision resistors with very small tolerances.

A suitable value for R2 can be calculated as:

$$R2 = \frac{R1 \times R_F}{R2 + R_F}$$

The same circuit with $R1 = R_F$ works with a voltage gain of -1, when it is known as a *phase inverter*.

The noninverting version of this circuit is shown in Fig. 16-8. Here the gain is given by:

$$gain = 1 + \frac{R_F}{R1}$$

Circuits of this type with low gains are known as *buffers*.

INTEGRATOR

A basic *integrator* circuit is shown in Fig. 16-9. Theoretically

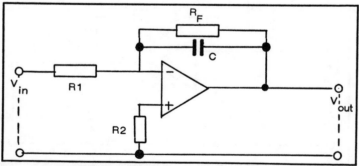

Fig. 16-9. Basic integrator circuit.

138

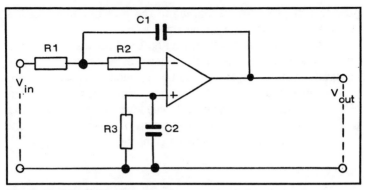

Fig. 16-10. A better integrator circuit.

this function is performed if the feedback resistor of previous circuits is replaced with a capacitor (C). For a practical circuit a resistor (R2) in parallel with C is needed to provide dc stability. This will have little effect at frequencies above about 30-40 Hz. Output is given by:

$$V_{out} = -\frac{1}{RC} \int_{o}^{t} V_{in} dt$$

There are snags to this circuit. One is finding a suitable value for R2 to hold the output level within suitable limits. It needs to be a high value to maintain good accuracy of integration, but low enough to avoid the limiting condition. Deciding a suitable value for R3 can be tricky, too, but here making R3 = R1 is usually a satisfactory solution. If you get things wrong there is a danger of destroying the op amp.

A rather better circuit which is safer in this respect is shown in Fig. 16-10. Here R3 can be made equal to R1 and R2 chosen to

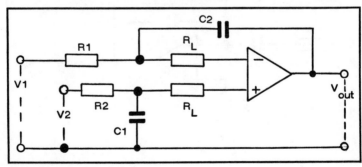

Fig. 16-11. Differential input integrator.

provide sufficient current limiting in the inverting input lead. A suitable value for most basic op amps would be 4.7 k ohm.

The integrator circuit shown in Fig. 16-11 carries this principle further to work with a differential input and at the same time provide integration without inversion (the two previous integrator circuits produce an inverted output). The output in this circuit is the integral of the differential input. However, it only gives correct integration if R1C1 = R2C2. In practice this means making R1 = R2 and C1 = C2 and using close tolerance components.

DIFFERENTIATOR

The basic *differentiator* circuit is similar to that of the integrator with the addition of a capacitor in the noninverting input feed (Fig. 16-12). Without capacitor C2 this circuit will still work but will be very noisy.

The performance of this circuit is given as:

$$V_{out} = R2C1 \; dV_{in}/dt$$

COMPARATOR

Special op-amps are available to work as comparators. These have faster slew rates and normally also incorporate an output transistor, calling for an additional voltage supply. They are not suitable for use as feedback amplifiers because they are difficult to stabilize against feedback oscillation.

A basic *comparator* circuit is shown in Fig. 16-13. R1 and R2 should be of equal value, using close tolerance resistors to minimize errors, although some offset error will inevitably be present. The

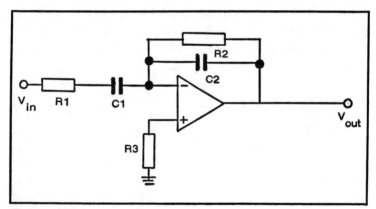

Fig. 16-12. Differentiator circuit.

140

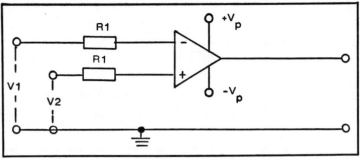

Fig. 16-13. Basic comparator circuit.

output can also usually benefit from smoothing, e.g., by an RC network (filter).

VOLTAGE FOLLOWER

Tying the two inputs of an op amp together as in Fig. 16-14 gives a voltage follower circuit where the output always follows the input:

$$V_{out} = V_{in}$$

The particular advantage of a voltage follower is that it offers a high input resistance with low input current and a very low output resistance. There are many applications of this type of circuit and a number of op amps are produced specifically as voltage followers with tied inputs.

VOLTAGE-TO-CURRENT CONVERTER

Connecting an op amp as shown in Fig. 16-15 will result in the same current flowing through the input resistance R1 and the feedback resistor or load impedance R2. The value of this current will be independent of the load, but proportional to the signal voltage. Its value will be directly proportional to V_{in}/R1. In practice this current will always be low because of the high input resistance presented by the noninverting terminal.

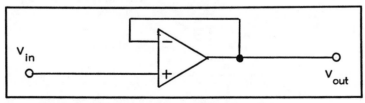

Fig. 16-14. Tying the two inputs together gives a voltage follower.

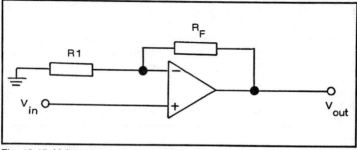

Fig. 16-15. Voltage-to-current converter circuit.

CURRENT-TO-VOLTAGE CONVERTER

In this case the connections are as shown in Fig. 16-16. This enables the input signal current to flow directly through the feedback resistor R2 giving an output voltage equal to $I_{in} \times R2$. Thus input *current* is converted into a proportional output *voltage*. No current flows through R2, the lower limit of current being established by the bias circuit generated at the inverting input. The capacitor shown in this circuit is added purely to reduce 'noise.'

OP AMP AS CURRENT SOURCE

Figure 16-17 shows an op-amp circuit for working as a current source, the value of the output current being given by:

$$I_{out} = V_{in} \times R3/R1 \times R5)$$

Resistor values should be selected so that:

$$R1 = R2$$
$$R3 = R4 + R5$$

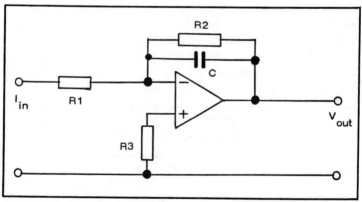

Fig. 16-16. Current-to-voltage converter circuit.

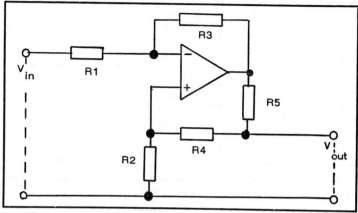

Fig. 16-17. Op amp connected as a current source.

SCHMITT TRIGGER

A basic circuit for a *Schmitt trigger* is shown in Fig. 16-18. This gives an output once a specific value of input has been reached, i.e., acts as a voltage 'trigger'. Equally it can operate as a dc voltage sensor, 'triggering' or switching on at a predetermined level of input voltage. The trigger or *trip* voltage is given by:

$$V_{trip} = -V_{out} \times \frac{R1}{R1 + R2}$$

Such a circuit has inherent 'backlash' or hysterisis equal to twice the trip voltage.

Schmitt trigger circuits are also derived from more complicated op amps together with external components such as resistors, diodes, and capacitors. It is difficult to justify their use since complete Schmitt triggers are available in IC form, avoiding the need for discrete components.

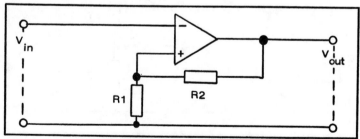

Fig. 16-18. Basic Schmitt trigger circuit using an op amp.

143

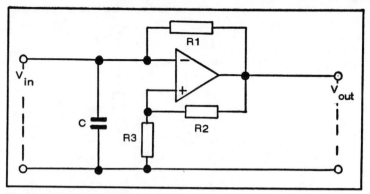

Fig. 16-19. Op-amp multivibrator circuit.

OP-AMP MULTIVIBRATORS

Op amps can also be made to work as multivibrators or oscillators. A basic circuit for a free running multivibrator is shown in Fig. 16-19. Feedback is controlled by R1, while capacitor C decouples the supply. The frequency of such a circuit is given by:

$$f = \frac{1}{2\,CR1\log_e \dfrac{2R3 + 1}{R2}}$$

A simple *monostable* multivibrator circuit is shown in Fig. 16-20. It is capable of being triggered by a square wave pulse. Input is 'protected' by a diode. Resistors R1 and R2 can be of similar value (say 1 k ohm). Circuit adjustment can be made via the variable resistor R4 (in the positive supply line, in series with a fixed

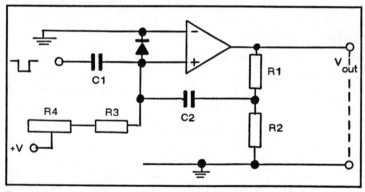

Fig. 16-20. Monostable multivibrator circuit.

144

resistor R3 to limit maximum current to + input). A suitable value for R3 would be 1 M ohm, with R3 0-5 M ohm or 0-10 M ohm. This circuit is a one-shot *pulse generator*.

OP-AMP WAVEFORM GENERATORS

Modifying this circuit as shown in Fig. 16-21 and shunting the output to ground by two zener diodes back to back limits the output to the zener voltages. A proportion of the output is fed back to the noninverting input via the potential divider formed by R2 and R3, the value of which is given by:

$$v = \frac{R3}{R2 + R3}$$

Thus the differential input voltage is:

$$V_{in} = V_c - v \times V_{out}$$

where V_c is the capacitor voltage.

If V_{in} at the noninverting terminal is positive, the V_{out} goes to the diode limited voltage $-V_{Z1}$. At the same time the capacitor C is charged to this voltage. When V_{in} at the noninverting terminal goes negative, then V_{out} goes to $+V_{Z2}$, with corresponding reverse charge of the capacitor. The result is a square-wave output with a time constant governed by the values of C and R4.

Specifically the time for one complete square wave is given by:

$$T = 2(C \times R4) \log_e \frac{1 + v}{1 - v}$$

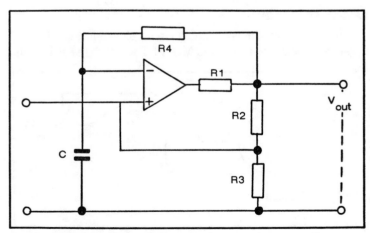

Fig. 16-21. Op-amp waveform generator.

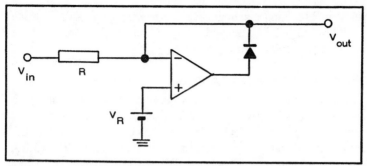

Fig. 16-22. Using an op amp as a clamp.

This simple circuit for a *square-wave generator* performs well up to about 10 kHz. Above that the slew rate of the op amp is liable to limit the slope of the output square wave.

OP AMPS AS CLAMPS

The circuit shown in Fig. 16-22 provides a smooth effective *clamp*, with the output equal to the voltage at the noninverting input. The output follows the input as V_{in} rises to V_R and is then clamped at V_R.

Two precautions should be observed. V_R should be chosen so that it is slightly larger than the maximum value of V_{in} otherwise the input stage will be saturated since there is no feedback. Also, if the diode is reversed biased a large voltage can appear between the inputs.

SAMPLE-AND-HOLD

In the simple sample-and-hold circuit shown in Fig. 16-23, a negative pulse at the gate of the MOSFET will turn the switch on.

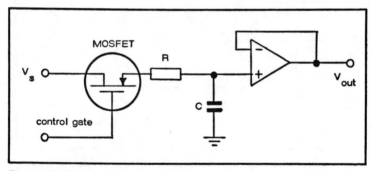

Fig. 16-23. Sample-and-hold circuit based on op amp.

The capacitor C will then change to the instantaneous value of the input voltage. In the absence of a pulse the MOSFET will switch off, when the capacitor is isolated from the output by the op amp. It will thus continue to hold the voltage across it. The resistor R is not an essential feature of this circuit and can be omitted with capacitor values of less than .05 μF capacity.

OP-AMP PARAMETERS

The important parameters for op-amps are:

☐ **Input offset voltage V_{os}.** This is defined as the input voltage required between the input pins to cause the output voltage to reach a level midway between the positive and negative supplies. A typical value is 1 mV.

☐ **Input bias current I_B.** This is the current taken by either of the input connections. Its value depends on the β of the input transistors and is variable with termination. A typical value is 100 nA.

☐ **Input offset current I_{os}.** This is the difference between the two input bias currents, typically about one third of the bias current but tending towards zero in a well balanced input circuit. A typical value is 10 nA.

☐ **Gain A.** This is the voltage gain at low frequencies from input to output. It can have a value of 100,000 or more in typical op amps, although dependent on frequency. A typical value would be 50,000.

☐ **Frequency Response.** This is defined by the break frequencies for each successive stage of amplification. It is affected by lag, but this can often be ignored for most applications.

☐ **Slew Rate.** This is the rate of swing at the output at its limiting value (i.e., with an infinitely fast input). It is specified in volts per second.

Chapter 17

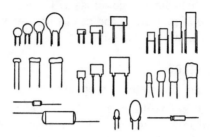

Inductances and Chokes

Coils carrying a current generate a magnetic field—the working principle of the electromagnet. If the value of the current changes, then the strength of the magnetic field will also change, inducing a back *emf* opposing the original current. A coil in an alternating current circuit, therefore, will have both its normal *dc* resistance plus some additional resistance due to inductance. This *ac* resistance is known as reactance (X_L), given by:

$$X_L = 2\pi f L$$

Where f is the frequency in Hz
and L is the inductance.

For inductance given in henrys, the reactance is in ohms. Practical values of inductance however are millihenrys (mH) or microhenrys (μH). Thus the above formula also gives X_L in ohms for:

frequency in kHz and inductance in mH.
frequency in MHz and inductance in μH.

A pure inductance is impossible as a practical component. Coil will necessarily have some resistance as well as inductance in an *ac* circuit, so is effectively equivalent to a (pure) resistance and (pure) inductance in series. Its total resistance in an *ac* circuit is then given by its *impedance* (Z), where:

$$Z = R^2 + X_L^2$$

The general design aim with inductors is to make R quite low so that its ultimate performance is basically dependent on its inductance value. For example, an inductor with a value of 1 μH may have a resistance of only 1 ohm; a 100 μH inductor a resistance of 2 ohms; and a 1 mH inductor a resistance of possibly 5 - 10 ohms. As a general rule the greater the value of inductance the higher the value of R inevitably has to be.

The particular value of an inductor is as a *choke*. Thinking only of its inductance as the significant value, this is directly proportional to the frequency of the signal. Thus at low frequencies its reactance is small, and at high frequencies its reactance is large. In other words, such a component can pass lower frequencies with little resistance, but offer high resistance to high frequencies, or choke such frequencies.

A particular example is the *radio frequency choke* (rfc). It is designed with low ohmic resistance but has high reactance at radio frequencies. It can thus pass *dc* with little resistance but will block radio frequency *ac* when both are present in the same circuit. In this respect it operates in the opposite way of a capacitor.

The characteristics of any *rfc* will vary with frequency. At high frequencies it will have characteristics similar to that of a parallel-resonant circuit; and at low frequencies characteristics similar to that of a series-resonant circuit. At intermediate frequencies it will have intermediate characteristics. The actual characteristics are relatively unimportant when an *rfc* is used for *series feed* because the *rf* voltage across the choke is negligible. If used for *parallel feed* (where the choke is shunted across a tank circuit), it must have sufficiently high impedance at the lowest frequencies and no series-resonance characteristics at the higher frequencies in order to reduce power absorption to a suitable level. Otherwise there will be a danger of the choke being overloaded and burnt out.

Chokes designed to maintain at least a critical value of inductance over the likely range of current likely to flow through them are called *swinging chokes*. They are used as input filters on power supplies to reduce 'ripple' or residual *ac* content. Chokes designed specifically for smoothing 'ripple' and having a substantially constant inductance, independent of changes in current, are known as *smoothing chokes.*

Chokes may be constructed with fixed values, or variable values. The former are wound on a fixed core; the latter on a bobbin with an adjustable (powdered iron) core and usually canned. For such components maximum inductance values are quoted—

typically up to 1 mH for miniature components; or up to 100 mH for larger commonly available sizes.

With chokes a maximum *dc* current value is often quoted. This does not necessarily imply that higher currents would damage the component. More usually this quoted value represents the dc current level which affects the quoted inductance value by more than 20 percent.

Transformers are also inductive components. See Chapter 26 for the effect of inductors combined with capacitors; also see Chapter 27.

Chapter 18

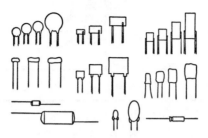

Transformers

The transformer is primarily a device for stepping up or stepping down an *ac* voltage. It is also widely used as a means of impedance matching, e.g., when coupling a power output into a low impedance load such as an *af* amplifier into a loudspeaker.

A transformer consists of two separate coils wound on a closed magnetic circuit. Provided the magnetic circuit has low reluctance (i.e., is a laminated iron core), very high efficiencies can be obtained, approaching ideal performance. The input side is termed the primary and the output side the secondary:

$$\text{input voltage} = \text{Np} \times \text{output voltage}$$

$$\text{input current} = \frac{\text{Ns}}{\text{Np}} \times \text{output current}$$

when Np = number of turns in primary coil
Ns = number of turns in secondary coil

This means that if a transformer is used to *step down* a voltage, there will be a *step up* in current in the same ratio, and if it gives a *step up* in voltage, there will be a *step down* in current in the same ratio.

Putting this into working formulas (see also Fig. 18-1):

$$V_{in} = n \times V_{out}$$

$$I_{in} = \frac{1}{n} \times I_{out}$$

$$\frac{V_{in}}{I_{in}} = \frac{1}{n^2} \times \frac{V_{out}}{I_{out}}$$

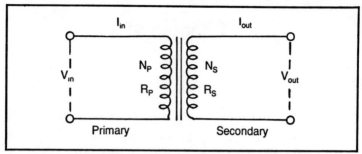

Fig. 18-1. Basic transformer notation.

where n is the turns ratio or N_P/N_S

Other significant parameters are the resistance of the primary winding R_P, the resistance of the secondary winding R_S, and the load resistance to be applied to the secondary output. Strictly speaking we should use *impedances* rather than resistance since the transformer works with *ac*, but calculations based on resistance values are much simpler and usually quite satisfactory.

The *actual* voltage across the primary coil will be a little less than V_{in}, because of the resistance of the primary coil. Similarly the *actual* output voltage will be a little less than V_{out} because of the resistance of the secondary coil. There will also be some losses in the magnetic circuit. These losses are normally small enough to be ignored, i.e., transformer performance is calculated purely on turns ratio.

There will, of course, be a further drop in output voltage across the load applied to it, depending on the resistance of this load. The *true* output power is thus related to the actual current flowing through the output load.

CALCULATIONS FOR STEP-UP OR STEP-DOWN AC VOLTAGE

Redesignating 'in' as the primary side (P) and 'out' as the secondary side (S), and designing to produce a required *output* voltage:

$$V_{out} = V_S = \frac{V_P}{n}$$

where n is the turns ratio:

This is normally worked as a formula for turns ratio:

$$n = \frac{V_P}{V_S}$$

Example: Find the turns ratio required for a transformer to change 110 Vac input to 40 Vac output:

$$n = \frac{110}{40}$$
$$= 2.75$$

Remember n is the ratio of the number of turns in the *primary* core to the number in the secondary coil. So in this example the primary coil needs to have 2.75 times as many turns as the secondary coil. If in doubt, remember the following:

☐ Stepping *down* a *voltage* means *more* turns on the primary.

☐ Stepping *up* a *voltage* means *less* turns on the primary.

CALCULATIONS FOR STEP-UP OR STEP-DOWN AC CURRENT

This follows exactly the same principle, but using the current formula:

$$I_{out} = n \times I_{in}$$
$$\text{or } I_S = n \times I_P$$
$$\text{Thus } n = \frac{I_S}{I_P}$$

Note that this reverses the 'turns ratio' rule given for voltage:

☐ Stepping *down* a *current* means *less* turns on the primary.

☐ Stepping *up* a *current* means *more* turns on the primary.

Example: Design a transformer to step up *ac* current from 1.2 amps to 5 amps

$$n = \frac{I_S}{I_P}$$
$$= \frac{5}{1.2}$$
$$= 4.17$$

Here the primary winding needs to have 4.17 times as many turns as the secondary winding.

IMPEDANCE MATCHING

The ideal relationship is:

$$Z_P = Z_S \times n^2$$

where Z_P is the impedance into the primary.

Z_s is the impedance of the load connected to the secondary.

Again this is normally worked as a solution for turns ratio:

$$n = \sqrt{\frac{Z_P}{Z_S}}$$

For a majority of applications we can use resistance values instead of impedances, when:

$$n = \sqrt{\frac{R_P}{R_S + R_L}}$$

when R_L is the resistance of the load.

Usually, too, it is not necessary to bother about the resistance of the secondary winding, which can be ignored (particularly as the load value may be nominal and variable, such as the impedance (resistance) of a loudspeaker).

We then have a very simple formula for impedance matching via a transformer:

$$n = \sqrt{\frac{R_P}{R_L}}$$

Specifically, the value of R_P *required* is determined first—e.g., to match the output requirements of a transistor stage. This is entered in the formula together with the known value of the output load to arrive at the *turns ratio* required.

THE AUTOTRANSFORMER

The *autotransformer* comprises a single winding on a magnetic core, with an intermediate tapping point. As shown in Fig. 18-2 the whole length of the coil becomes the primary with the secondary tapped off from some point along it. This is a voltage *step-down* autotransformer or current step-up autotransformer. It can also be

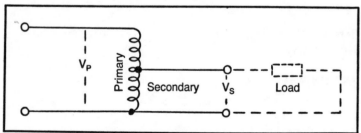

Fig. 18-2. Voltage step-down autotransformer.

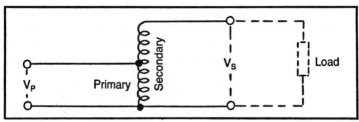

Fig. 18-3. Voltage step-up autotransformer.

worked in the opposite mode for voltage *step-up* (or current step-down), by reversing the roles of the primary and secondary, (Fig. 8-3).

The same transformer rules apply for performance calculation, remembering that:

☐ In the voltage *step-down* autotransformer the number of turns in the primary is the total number of turns in the coil; and the number of secondary turns is that from the tapping point down to the common connection.

☐ In the voltage *step-up* autotransformer, the secondary is comprised of the full number of turns and the primary of the number of turns from the tapping point down.

There are two specific advantages of an autotransformer:

☐ With only one winding required the amount of copper is reduced (reducing winding losses). However, this is only significant when the turns ratio is low (about 10 percent saving for a turns ratio of 10, rising to 75 percent with a turns ratio of 1.3).

☐ Unlike a potential divider an autotransformer can be used to step-up a voltage as well as step-down a voltage, simply by changing over primary and secondary connections.

On the disadvantage side it is not usually as efficient as a conventional transformer with separate primary and secondary coils. Also the two windings are not separated electrically. Thus if one side is carrying a high voltage, a breakdown could result in this voltage being connected directly to the other side. If the auto-transformer was being used to step-down 110 volts to, say 10 volts, for example, a failure in the length of coil joining the secondary could result in 110 volts being fed directly into the 10 volt output side.

I-F TRANSFORMERS

I-f or *intermediate frequency transformers* are specialized components used in superhet radio circuits. They are basically *tuned circuits* which provide coupling between the stages of the *i-f* amplifier circuit and normally consist of a canned coil with an iron dust core for inductive tuning.

In its simplest form the primary is associated with a capacitor of fixed value, forming a circuit which can be tuned via the iron dust core to resonate at the intermediate frequency (i.e., usually 455 kHz or 470 kHz, or possibly 1.6 MHz). This is a single tuned i-f transformer. Alternatively, both primary and secondary coils are associated with capacitors to form separate (tunable) tuned circuits. This is a double-tuned i-f transformer.

Choice of type depends on the amplifier circuit involved. If only a single i-f stage is used, then a double-tuned i-f transformer is more or less essential. With two or more i-f stages, single-tuned i-f transformers may be quite satisfactory for interstage coupling.

The other main requirement is that the input of the first *i-f* transformer must correctly match the output load impedance required by the preceding transistor stage, and the output from the secondary match the input impedance of the next stage; and so on. Mismatching will materially reduce the gain through the i-f amplifier stage(s).

I-f transformers are complete units in themselves, i.e., incorporate associated capacitor(s) within the can. The metal can itself also performs an important function (that of screening). They are widely available in subminiature sizes (approx. 10 mm square by 12 mm high) and miniature sizes (approx. 13.5 mm square and 17.5 mm high) with output pins designed for printed circuit board or chassis mounting. Pins 1, 2, and 3 are normally the input side (pin 2, if present being a center-tap to the primary). Pins 4 and 5 are the output side; or 4, 5, and 6 of the secondary is also tapped (pin 5).

I-f transformers are sometimes replaced by *transfilters*. These are solid-state piezoelectric devices comprising, basically, a crystal clamped between two electrodes. A transfilter will resonate at one frequency only, i.e., is 'fixed' tuned'. Their use can simplify alignment problems on superhet radio receivers although they do have certain limitations. For this reason they have not generally replaced *i-f* transformers.

Chapter 19

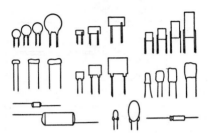

Contacts

Modern electronic circuits aim at eliminating components as far as possible, e.g., replacing relays by solid-state switching devices. However, there is still a requirement for spring-loaded contacts in many applications and the following will provide a useful design reference.

CONTACT SPRINGS

Contact springs (Fig. 19-1) normally are flat springs and the deflection and stress are related by the following formulas:

$$\text{deflection (d)} \quad \frac{4PL^3}{bt^3E}$$

$$\text{stress (S)} = \quad \frac{6PL}{bt^2}$$

where P = load or spring pressure
L = length of spring
b = width of spring
t = thickness of spring
E = modulus of elasticity for spring material

Typical values of moduli of elasticity for the usual contact spring materials in inch units are:

Brass* 15×10^6
Phosphor bronze (spring temper) $16 - 17 \times 10^6$

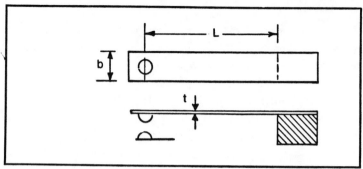

Fig. 19-1. Basic contact spring geometry.

| Beryllium copper (heat treated) | $16 - 19 \times 10^6$ |
| Nickel silver (spring temper) | $10 - 20 \times 10^6$ |

*Not normally selected as a contact material, but may be used in simple, inexpensive designs.

Maximum contact pressure obtainable from a flat spring can be calculated from:

$$P_{max} = \frac{bt^2S_p}{6L}$$

where S_p is the maximum permissible material stress. This can be taken as 80 percent of the limit of proportionality of the material:

Phosphor bronze—24-25.5 tons/sq.in.
Beryllium copper—37-40 tons/sq.in.
Nickel silver—16-18.4 tons/sq.in.

In practice spring pressure is normally adjusted by bending the spring, i.e., the setting of the unloaded position of the spring is such that the movement (deflection) produces the required spring force. Spring design thus calls for suitable selection of L, t, and b in order to arrive at a suitable maximum value of spring pressure likely to be required, for a given spring material.

CONTACT PRESSURE

Contact pressure provides the closing force required to complete a suitable contact for current flow, if necessary sufficient to rupture any tarnish film on the contact surfaces. An increase of contact pressure over this minimum force required will then, as a general rule, reduce contact resistance and increase the maximum current which can be carried by the contacts.

Use Fig. 19-2 as a design chart for determining lead diameter and closing force for simple contacts. Maximum permissible working stress in the spring material should not normally be taken as greater than 80 percent of the limit of proportionality. P represents the maximum available contact pressure for that particular spring of width B and thickness T. The nomogram scales enable pressure to be read in pounds or grams. Figures for P obtained from the nomogram refer to a spring length of 1 inch. For other spring lengths, the answer obtained must be divided by the length of the spring in inches.

Example: Find the maximum contact pressure obtainable from a beryllium-copper leaf spring 2½ inches long by 0.025 inches thick and 3/16th inches wide. Enter values on the Fw and B scales first. From the point established on the reference line, project across the value of T. Read off contact pressure P (790 grams or 1.74 lb). To obtain a correct value for a 2½ inch long spring this value must be divided by the spring length (790 ÷ 2.5 = 317 grams or 1.74 ÷ 2.5 = 0.697 lb. Values of limit of proportionality for commonly used leaf contact spring materials are:

> Beryllium copper—100,000 - 110,000 psi
> Nickel-silver—45,000 - 50,000 psi
> Phosphor-bronze—67,500 - 72,000 psi

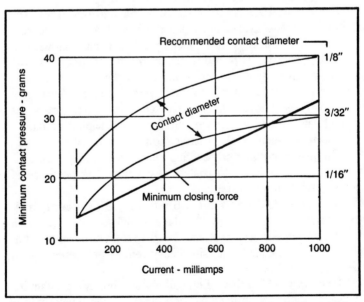

Fig. 19-2. Graph for determining suitable contact size and closing force (spring pressure required).

CONTACT MATERIALS

Silver. The most generally useful of all contact materials. It has higher electrical and thermal conductivities than any other metal, does not oxidize and gives a low contact resistance even with light pressures. Unsuitable for extremely sensitive contacts that are to carry small currents at low voltages, or that are infrequently operated, particularly when atmosphere contains sulphur compounds, which form a sulphide film on the contact faces.

Electrodeposited silver. Hard, dense and firmly adherent. The process of electro-deposition offers the means by which thin facings may be applied economically, and thick facings without softening of the backing material.

Platinum. Suitable for applications where contact pressures are low and where reliability is of the greatest importance. Completely immune to oxidation or tarnishing at any temperature.

Palladium. A useful alternative to platinum, in suitable conditions, but less resistant to electrical erosion.

Rhodium. Applied as an electrodeposit, is extremely hard and completely immune to tarnishing. Gives a low and stable contact resistance, but will not withstand sparking or arcing.

10 percent gold-silver. Has slightly greater resistance to tarnishing than pure silver and is slightly harder.

5 percent, 10 percent, 20 percent palladium-silver. The addition of increasing amounts of palladium to silver gives progressively greater resistance to tarnishing and to mechanical wear.

7½ percent, 10 percent, 20 percent, 50 percent copper-silver. Resistance to tarnishing diminishes with increasing copper content, though hardness and resistance to mechanical wear increase. These alloys give good service when contact pressures are high, and when the contacts operate with a wiping action.

Cadmium-silver. Alternatives to the copper-silver series.

Cadmium-copper-silver. Superior to resistance to tarnishing, but greatly inferior to contact resistance. The presence of cadmium tends to quench arcing.

10 percent, 20 percent, 25 percent, 30 percent, iridium-platinum. These alloys combine hardness, increasing with the iridium content, with complete freedom from tarnishing. Suitable for sensitive equipment using heavy contact pressures at high operating speeds.

Irru and 540 alloy. Both alloys are highly resistant to mechanical deformation and the effects of arc erosion. They are free from tarnish and oxidation at ordinary temperatures.

40 percent silver-palladium. Complete resistance to sulphide film formation together with maximum hardness and resistance to wear of the silver-palladium series. An alternative to platinum in certain conditions.

40 percent copper-palladium. Similar characteristics to 40 percent silver-palladium, with slightly less resistance to tarnish but greater resistance to material transfer.

30 percent silver-gold. Substitutes for platinum in sensitive apparatus, though they are inferior in resistance to electrical wear.

Platinum-silver-gold. Extremely useful for rubbing or sliding contacts in the form of wire or strip.

Copper-silver-gold. Has excellent spring properties and is completely resistant to tarnishing.

Tungsten. Very hard, with good resistance to welding and material transfer. High contact resistance prohibits its use in low voltage circuits, or with low contact pressures.

Copper-tungsten IW3, 3W3, 30W3. Combine the high conductivity and relatively low contact resistance of copper with the hardness and resistance to arc erosion of tungsten.

Silver-tungsten 50S, 35S, 20S. Have properties similar to those of the copper-tungsten group, but are superior in contact resistance and in resistance to oxidation.

Silver-tungsten carbide G13, G14. These materials have great hardness.

Silver-molybdenum G17, G18. They will withstand electrical erosion and mechanical wear in heavy duty relays in which contact pressures are above normal.

Silver-cadmium oxide D54, Silver-nickel D56. Give considerably greater resistance to erosion than fine silver, together with a low contact resistance. Used as alternatives to silver in dc circuits.

Silver-graphite D58. The small percentage of graphite in this material gives it self-lubricating properties and protection against welding.

Table 19-1 will help in the selection of contact materials.

FORMS OF CONTACTS

Solid Headed Rivet Contacts. These are manufactured by cold forging, the most economical method of mass production.

Solid Turned Rivet Contacts. Such contacts are manufactured when the quantity ordered is insufficient to justify setting up a heading machine (usually 5000), when the design of the contacts is

Table 19-1. Contact Material Selection Guide.

Type of Apparatus	Conditions	Contact Material
Sensitive relays and instruments	Current and voltage small. Very light pressure. Complete freedom from surface films on contacts essential in order that contact resistance remains low.	Copper-palladium Platinum-silver-gold Platinum Iridium-platinum Rhodium
Telephone relays	Circuit conditions unlikely to cause rapid arc erosion or material transfer.	Silver Platinum-silver-gold
	For more severe duty	Palladium Platinum
Switches handling only radio-frequency or audio-frequency currents	Maintenance of low and constant contact resistance essential.	Palladium Platinum Rhodium
	Where a heavy wiping action maintains clean contact surfaces	Copper-silver Silver
Sliding contacts	In radio-frequency circuits.	Silver/silver Rhodium/silver Rhodium/rhodium
	In dc and ac circuits	Silver/silver Copper-silver-gold/silver Copper-silver-gold/rhodium
Small dc signalling devices	Moderate current and voltage. Sufficient arcing to break down slight surface film, but material transfer troubles unlikely to be an important feature.	Silver Palladium-silver Gold-silver Copper-palladium
Industrial heavy duty telephone-type relays	Current moderate on mains voltages, and inductive conditions probable. Frequency of operation moderately high.	Silver Silver-nickel Elkonite D56 Silver-cadmium oxide Elkonite D54
	Arcing conditions more severe, but frequency of operation reduced and contact pressure increased.	Silver-molybdenum Elkonites G17 and G18 Silver-tungsten carbide Elkonites G13 and G14
Thermostats	Moderate to heavy currents on mains voltages ac, high ambient temperatures, slow speed of make	Silver Palladium-silver
	As above but operating on dc	Silver Silver-nickel Elkonite D56 Silver-cadmium oxide Elkonite D54
Domestic switches	Moderate currents on domestic ac voltages	Silver
Magnetos	Alternating current type. Hydro carbon vapors not present in contact-breaker in any quantity.	Ruthenium-platinum Iridium-platinum Iridium-ruthenium-platinum ·
	Uni-directional current and hydrocarbon vapors present in quantity.	Tungsten

162

Table 19-1. Contact Material Selection Guide. (Continued from page 162.)

Voltage regulators	Moderate currents on battery voltages. Inductive currents. Hovering action of contacts.	Cadmium silver Silver Silver-nickel Elkonite Gold-silver Silver-tungsten Elkonite
Contactor-starters	Moderate to heavy currents on voltages up to 550 ac where frequency of operation is high.	Silver
	Heavy duty, and where frequency of operation is not high.	Silver-tungsten Elkonite
Industrial contactors	Where copper fails owing to high contact resistance caused by long idle periods in either closed or open position, or by overheating arising from high frequency of operation.	Silver Cadmium-copper-silver Silver-nickel Elkonite
	Where auxiliary arcing contacts are employed.	Silver for main contacts Silver-tungsten Elkonite for arcing contacts
Motor vehicle starter switches	Heavy currents on battery voltages. Inductive circuits. Occasional operation.	Silver Silver-nickel Elkonite Silver-tungsten Elkonite
Small circuit-breakers on domestic or battery voltages	No auxiliary arcing contacts employed.	Silver Silver-nickel Elkonite Silver-molybdenum Elkonite Silver-tungsten Elkonite
	Auxiliary arcing contacts used	Silver Silver-nickel Elkonite for main contacts Silver-tungsten Elkonite for arcing contacts
Air circuit-breakers	Main contacts, required to maintain minimum contact resistance to restrict temperature rise.	Silver Silver-nickel Elkonite
	Arcing contacts, required to resist erosion and to maintain sufficiently low contact resistance to protect main contacts at the instant of opening.	Silver-tungsten Elkonite
Oil circuit breakers	Main contacts, where minimum contact resistance is required and protection is afforded by the arcing contacts.	Silver Silver-nickel Elkonite
	Arcing contacts, required to resist and to maintain sufficiently low contact resistance to protect main contacts at the instant of opening	Copper-tungsten Elkonite
Air-blast circuit-breakers	Main contacts, required to maintain minimum contact resistance but to withstand some arcing	Silver-nickel Elkonite
	Arching contacts, required to resist arcing only	Copper-tungsten Elkonite

unsuitable for heading, or when the contact material is not ductile enough for cold forging.

Composite Turned Rivet Contacts. These are similar to solid turned rivet contacts, except that the contact material is brazed or welded to a base metal backing either for economy or because the contact material is unsuitable for riveting.

Tubular Rivet Contacts. A special form of mass produced headed composite contact is made in which a facing of silver is bonded to a copper backing with a hollow shank.

'Optecon' Silver Faced Rivet Contacts. This is another type of mass produced silver on copper composite contact. For quantities of more than 100,000 they are the cheapest form of silver faced rivet available.

Composite Button Contacts. These coined contacts consist of a facing of silver on nickel plated mild steel or copper nickel disc. The backing has a projection for easy resistance welding to the contact support.

Solid Button Contacts. These are silver buttons which have a projection on one face to facilitate resistance welding onto copper based backings.

Inlaid Contact Bi-Metal. This strip material consists of base metal inlaid or faced on one or both sides with a contact material.

Onlaid Contact Bi-Metal. This form is similar to inlaid contact bi-metal, but the contact material stands out from the base metal to which it is bonded.

Rod and Wire. These forms are available in any required diameter for the manufacture of contacts.

Strip. Rolled strip is supplied in a wide range of thicknesses for contact facings and special contact parts.

D-Section Rod. D-shaped rod in silver or copper-silver alloy is supplied in random lengths or cut into short pieces ready for brazing onto contact backings.

Electrodeposits. Electrodeposits of certain metals are used when it is uneconomical or impossible to produce the contact surface in any other way.

Facings and Inserts. Ready for brazing to backings.

Bimetal clad wires. The contact material being in the form of a thin surface coating on copper-alloy wire.

Chapter 20

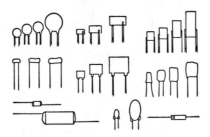

Transistor Switches

To operate as a switch the transistor is employed in a circuit which allows the transistor to be saturated or 'bottomed.' A basic circuit for achieving this is shown in Fig. 20-1, which works in the following way. As the collector-emitter voltage V_{ce} falls and approaches zero, the ratio of the charge in I_c to the charge in I_B (or β) falls and finally becomes $\beta = 0$. In this condition the transistor is saturated and any further increase in I_B has no effect on I_c / β_L, where β_L is the large signal current gain from 0 to the design value of I_c. Specifically this implies that:

$$R_B \text{ should be less than } R_L$$

SWITCH-OFF CIRCUIT

A basic switch-off circuit is shown in Fig. 20-2. In order to reduce the load current to zero (i.e., switch 'off') the base of the transistor must be more negative than the turn-on potential for the transistor. For an npn transistor the base must be more negative than the emitter and collector. For a pnp transistor the base must be more positive than the emitter and collector.

In the cut-off condition there is a current flow I_{cbo} out of the collector, and a current flow I_{cbo}, out of the emitter. I_{cbo} flowing into R_L gives a voltage drop equal to $I_{cbo} \times R_L$ across the load.

SWITCH-ON CIRCUIT

Here the circuit has to be designed so that the transistor base

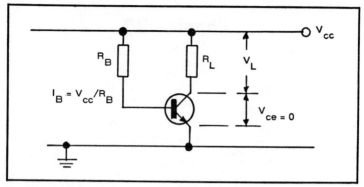

Fig. 20-1. Basic circuit for a transistor switch.

conditions change from a known base current I_B to a different working condition giving a known base voltage V_B with a directly coupled input (Fig. 20-3). The main requirement is that the potential V1 recorded on the positive swing is considerably greater than V_{be} so that the base current I_B is almost equal to $V1/R_B$.

Where it is desirable to eliminate *dc* coupling, capacitor coupling is generally not satisfactory. This is because the capacitor will tend to build up a charge to a value of $V1 - V_{be}$ so that the transistor does not switch on again. This can be overcome by using a diode in the grounded emitter circuit to enable the capacitor to charge *and* discharge on each input cycle.

An alternative solution is to use transformer coupling (Fig. 20-4). The particular limitation here is the low efficiency of transformers at very low frequencies, resulting in low base currents. As a result, transformer coupling is normally only considered for high-frequency circuits.

BIDIRECTIONAL TRANSISTOR SWITCH

The 'saturation' circuit shown in Fig. 20-5 differs in that the

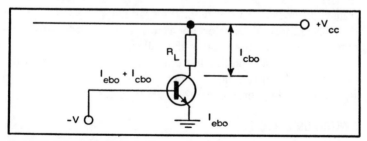

Fig. 20-2. Basic switch-off circuit.

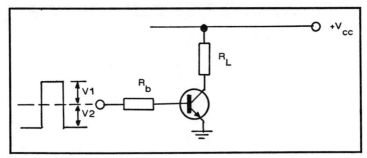

Fig. 20-3. Basic switch-on circuit.

transistor in the saturated condition acts as a short circuit to currents passing through R_E in either direction, provided $+V/R_B$ is greater than $-V/R_E$. In other words, V_{ce} will remain zero even when $-V$ is changed to $+V$.

A design point to remember here is that most transistors are not 'symmetrical', i.e., β is different in the forward and reverse directions. It is the β value in the required direction which must be used for calculation of R_E (and subsequent determination of R_B). Similar considerations apply as regards couplings as previously, except that emitter and collector are interchanged.

MECHANICAL SWITCH EQUIVALENT

A basic transistor circuit capable of performing the same function as a mechanical switch or relay is shown in Fig. 20-6. Design procedure is as follows:

(i) Choose type of transistor, pnp or npn, according to the 'switching polarity' required.

(ii) Determine the load currents $= V/R_L$ where V is the supply voltage to be mounted on and off and R_L is the load.

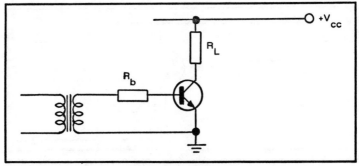

Fig. 20-4. Transistor switch with transformer coupling.

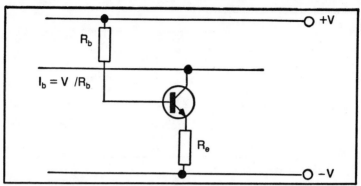

Fig. 20-5. Bi-directional transistor switch.

(iii) Select a suitable transistor having a large current gain L at $I_c = V/R_L$. Note: it should also have low leakage current.

(iv) Calculate the supply current to turn the transistor on at $2 \times V/R_L$.

(v) Check that V_{be} services at least reach zero, to turn the transistor off for this make supply voltage V_{cc} 10 × V_{be} at saturation.

(vi) Find value of R_B from the base current.

(vii) Check cut-off conditions.

Actual base potential $V_b = V_{cc} - (R_s \times (1_{cbo} + 1_{ebo})$

Check that V_b is zero or positive taking transistor characteristics at maximum operating temperature—say 80°C.

(viii) Check transistor power rating required: maximum power $= I_L V_{ce}$ (sat) $+ I_b \times V_{be}$

Example: Design a transistor switch to switch on a 30 volt *dc* supply connected to a 60 ohm load.

(i) Consideration of 'switching polarity' in the circuit of Fig. 20-6 requires that an npn transistor be used.

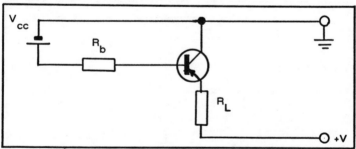

Fig. 20-6. Mechanical switch or relay equivalent circuit using a transistor as a switch.

(ii) Load current = V/R_L
$$= 30/60$$
$$= 0.5 \text{ amp}$$

(iii) The transistor must carry at least 0.5 amps—say 1 amp for safety, at 30 volts.

(iv) Suppose the transistor chosen has a L of 30 at 0.5 amp. For the saturated condition I_B must exceed

$$2 \times V/ L R_L$$
$$= 2 \times 30/30 \times 60$$
$$= 33 \text{ mA}$$

(v) V_{be} at saturation for the chosen transistor, is say 0.75 volts. Supply voltage sequence is thus $10 \times 0.75 = 7.5$ volts.

(vi) Base current = $(7.5 - 0.75)/R_B$

$$+ 33 \text{ mA (from step iv)}$$

thus $R_B = \dfrac{6.75}{.033}$

$$= 204.5 \text{ ohms}$$

Take the next *higher* preferred number:

$$R_B = 220 \text{ ohms}$$

(vii) Actual base potential $V_b = 7.9 - (R_B \times (I_{cbo} + I_{ebo}))$ Say transistor characteristics at 50°C are:

$$I_{cbo} = 10 \text{ mA}$$
$$I_{ebo} = 10 \text{ mA}$$

$$V_b = 7.5 - 220 \times .020$$
$$= 7.5 - 4.4$$
$$= 3.1$$

This is satisfactory. In fact V_{cc} could be reduced in value to, say 4.5 volts (if you do recalculate R_B and re-check.)

(viii) Maximum power = $I_L V_{ce(sat)} + I_b V_{be}$

$$= 0.5 \times 0.75 + .033 \times 0.75$$
$$= 0.75 \times 0.533$$
$$= 400 \text{ mW}$$

Note: the mean power to be dissipated by the transistor will depend on the time on/time off or mark-space ratio. For 'on' status of a few milliseconds this power figure will represent an adequate rating for the transistor. For extended 'on' periods, the power rating needs to be that of the load, i.e., $30 \times 0.5 = 15$ volts.

Chapter 21

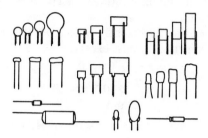

Analog Switches

The availability of *dc* voltage-controlled fully-electronic analog switches in IC form offers numerous functional advantages over mechanical switches. The most important being:

☐ The electronic analog signal switch can be positioned at a place in the circuit suitable for the wiring. The critical leads through which the signals are passed to mechanical switches (on the front panel) being replaced by control leads which only convey dc voltage signals. No long screened signal wires are thus required, which under certain circumstances can adversely affect the frequency response because of their high line capacitance and designers will have much greater freedom of choice of equipment design than with the mechanical signal switches.

☐ The electronic switches, when suitably designed, may be triggered by mechanical switch contacts as well as by electrical control signals supplying, for example, a circuit for touch-sensitive contacts or remote control.

☐ Repairs may be expected to be greatly reduced because of the absence of mechanical parts which are subject to wear.

☐ The use of electronic switches finally simplifies the introduction of modular designs.

Electronic switches of this type are called analog switches because the active stages of the circuits in the signal path must in the first instance transfer analog signals, regardless of whether the output is digital or not. Because of their importance they are described here at some length. Also, since such devices are individual designs, performance characteristics are specific to individual types. The following relates to the TDA 1028 and TDA 1029 switches developed by Valvo, with description by Mullard. The TDA 1028 contains two independently controllable double-pole switches with two switch settings each (two switches with 2 × 2 contacts each), whereas the TDA 1029 contains one double-pole switch with four switch settings each (one switch with 2 × 4 contacts).

These electronic circuits do not only transfer analog af signals, but also analog dc voltage signals with high accuracy, thus making them also suitable for use as switches in measuring equipment, e.g., voltmeters.

In the following section the operation and properties of the two analog switches will be described in more detail and also some typical uses of these circuits will be given.

CIRCUIT DESCRIPTIONS

The simplified theoretical circuit diagrams in Figs. 21-1 and 21-2 show clearly that the circuit TDA 1028 contains two double-

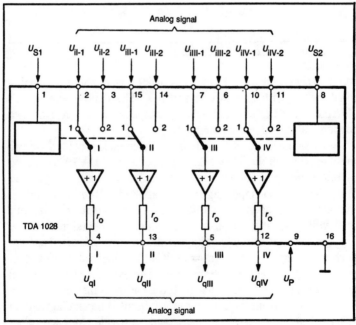

Fig. 21-1. Theoretical diagram TDA 1028.

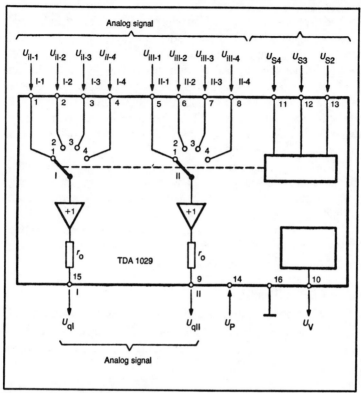

Fig. 21-2. Theoretical diagram TDA 1029.

pole switches and the circuit TDA 1029 one double-pole analog signal switch with four switch settings. The double-pole switch referred to here is a switch which is capable of transferring two independent signals simultaneously.

Each of the analog signals to be transferred is provided with an amplifier, (four in the TDA 1028 and two in the TDA 1029). These amplifiers are operational amplifiers employing full feedback in which the external input signals are applied to the noninverting inputs. Because of the strong feedback, the voltage gain is very small within the large permissible modulation range (signal voltage $U_{imm\ max} = 16$ V at a supply voltage of 20 V. As a safeguard against short-circuiting a resistor ($r_o = 400\Omega$) is connected internally in series with the output of each amplifier. The individual transmission channel of the circuits are designated by Roman figures.

As is seen from the slightly more detailed block diagrams in Figs. 21-3 and 21-4, a special input stage is provided for each signal

172

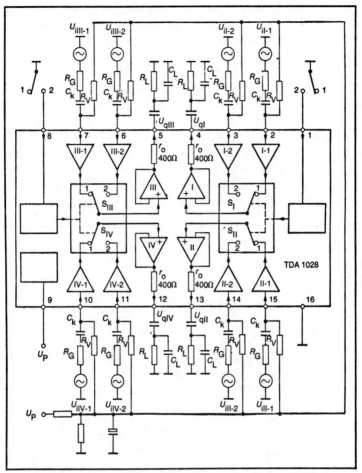

Fig. 21-3. Block diagram of TDA 1028 with example of external wiring.

input of the circuit. Figure 21-5 shows that two diodes are connected internally to each signal input, one between input and earth and the other between input and positive supply voltage U_p. The diodes are polarized in such a way that they are blocked when the value of the input voltage is between earth potential and positive supply voltage. As soon as the input voltage moves to outside this range, one of the diodes will become conductive, so that by means of suitable external current limiting resistors connected in series with the inputs the input circuits can be safeguarded against overvoltages.

The actual signal change is affected by changing over the

internal supply direct currents. Only the relevant preamplifier fully supplied with direct currents is able to operate and is switched in for the intelligence signals. The signal paths to the other preamplifier are blocked to the intelligence signals.

To reduce surges in the output, *dc* voltages go to a minimum when switching takes place. The circuit has been so designed that the direct currents to all the circuit inputs are equal to each other within narrow tolerances. When the circuit is used in practical applications, it is only necessary to ensure that the effective *dc* resistances inserted in the external input circuit of a switch are equal in value. The necessary switching of the supply direct current is effected in each double-pole switch with the aid of a control stage which is triggered by external switching signals and is also known

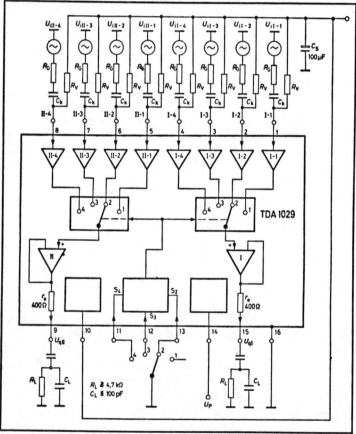

Fig. 21-4. Block diagram of TDA 1029 with example of external wiring.

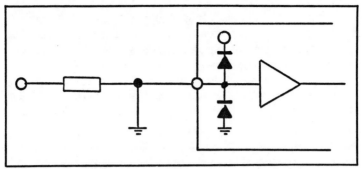

Fig. 21-5. Position of protective diodes at the analog signal inputs.

briefly as 'switch control.' In the TDA 1028 with two switch settings the two switch controls each have a control input. The internal switches are in one setting when the switching voltage to the associated control input < 2.1 V (LOW-state), i.e., when the input is connected to earth. The required control current flowing from the circuit is then less than 200 μA.

The other setting of the internal switch is obtained when the HIGH state occurs at the control input, which is the case when either the switching voltage at the control input exceeds 3.3 V or the control current flowing from the circuit drops below 0.1 μA. The HIGH state is thus certainly obtained when the control input is open-circuited, provided that the external resistance between the control input and earth > 33 MΩ.

In the TDA 1029 with four settings on each of the internal switches, the switch control has three control inputs. The switch setting governed by the states at the control inputs is as follows:

Switch setting	Terminal	11	12	13
	Voltage U	U_{S4}	U_{S3}	U_{S2}
1		H	H	H
2		H	H	L
3		H	L	X
4		L	X	X

where L = LOW-state, H = HIGH-state and X = any state. The conditions under which the LOW-state and the HIGH-state occur here are the same as those in the TDA 1028.

By means of an internal logic circuit, the switch setting always moves only to the LOW-state of any control input with the lowest terminal number, thus ensuring that no undefined switch settings and hence undesirable superimpositions or distortions can occur in

175

the output signal in the case of signals being applied simultaneously. By a suitable choice of control inputs, a sequential switch triggering system may be obtained, e.g., for functions to be switched on with priority over other functions.

In the analog signal circuits TDA 1028 and TDA 1029, the switching thresholds are internally stabilized and the limit values of the thresholds thus apply throughout the whole permissible range of supply voltages. In both types of circuits the switching input voltage may be equal to the supply voltage, so that it is very easy with the aid of lamps connected between the switching inputs and the supply voltage terminal, to indicate the switching state. Also connected internally between the inputs and the earth terminal are 16 diodes which protect the circuits against too high negative switching voltages, provided that the switching currents are inadequately limited.

The analog signal switches are normally operated with a positive supply voltage connected between earth and the supply voltage terminal. The amplifier is then driven by voltages within a positive range of values, so that the af signal voltages to be transmitted must be superimposed by a suitable positive bias. The possible ways in which this can be done will be discussed later. A suitable positive bias at which the drive limits are symmetrical about the operating point is provided by the internal bias supply source of the TDA 1029. For the TDA 1028, the bias must be obtained either from an external source or from the internal source of the TDA 1029 where this is used in the same circuit.

It should be pointed out here that, provided that the absolute limit values are observed, it is obviously also possible to operate the analog signal switches with a supply voltage balanced to the external earth. Finally, the circuits TDA 1028 and TDA 1029 contain stages in which the internal auxiliary voltages and currents required to operate the circuits are derived from the external supply voltages U_p. Several of the internal operating currents are stabilized.

SUPPLY VOLTAGE, QUIESCENT SUPPLY CURRENT, AND DISSIPATION

The recommended supply voltage U_p to which the data to be given below refers is 20 V, unless otherwise stated. The range of voltages which can be used in practice is limited on the one hand by the maximum permissible supply voltage $U_{Pmax} = 23$ V and on the other, by the specified lower limit of this voltage $U_{Pmin} = 6$ V at which the circuit still operates satisfactorily. The effective modulation range of the circuits obviously decreases with the supply voltage.

The supply current I_p, without the load resistors being connected to the outputs, amounts to approximately 2.5 mA for the TDA 1028 and approximately 4 mA for the TDA 1029. The dissipation P_{tot} of the circuit can be estimated most adequately from the equation:

$$P_{tot} = U_p I_p + \sum_{v=1}^{n} U_{Qv} I_{Qv}$$

where the dc voltages U_{Qv} and the direct currents I_{Qv} refer to the four output terminals 4, 5, 12, and 13 in the TDA 1028 and to the output terminals 9 and 15 in the TDA 1029. The currents have a positive sign when they flow into the circuit. The dissipation may not exceed a value of $P_{tot\,max} = 800$ mW at 80°C.

INPUT VOLTAGES AND INPUT
CURRENT AT THE ANALOG SIGNAL INPUTS

The input voltage U_I is here the voltage between the external analog signal inputs and the earth terminal 16. The diodes connected internally to the signal inputs limit the input voltage range to values between $U_{Imin} = -U_{BE}$ and $U_{Imax} = U_P + U_{BE}$. At input stages U_I falling outside the range $U_I = -0.5$ V to $U_I = U_P$ it is naturally necessary to ensure by limiting the current externally that the average input current remains below a value of $I_{IAVmax} = 20$ mA in order to protect the diodes. Single, sufficiently short input current pulses may well have an amplitude which is markedly greater than the maximum average input current.

The operating range of the input voltages in which the transfer characteristic of the amplifier is largely linear is less than the permissible input voltage range and is between approximately +3 V and $U_P = 1$ V, i.e., at $U_P = 20$ V between + 3 V and + 19 V. This drive range corresponds roughly to the common mode voltage range of the internal operational amplifier.

From the given operating range of the input voltages, the maximum peak-to-peak value of the signal input voltage is obtained immediately, i.e., $U_{i\,p-p\,max} = U_P - 4$ V. A prerequisite for the largely undistorted transfer of such a large signal is, of course, that the signal voltage is superimposed by a suitable bias U_V for obtaining the right operating point. If the peak values of the signal voltage are arranged symmetrically about earth potential, as is the case with *af* signals in the broadcasting range, the supply voltage U_V will be in the center of the input voltage operating range, hence at $U_V = 0.5\,U_P = 1$ V.

Assuming that these conditions are met, a sinusoidal input signal with a maximum rms value of $U_{i\,rms\,max} = U_{i\,p\text{-}p\,max}$ can thus be transferred undistorted. At a supply voltage $U_P = 20$ V a typical value obtained in practice for the input voltage at which the distortion of the output voltage in the mean frequency range is exactly 1 percent will be $U_{i\,rms\,max} = 6$ V.

INPUT QUIESCENT CURRENT, INPUT
OFFSET CURRENTS, AND INPUT OFFSET VOLTAGES

To operate the circuit the base dc currents must be applied to the input transistors across the external signal inputs. A typical value for these base currents which are also described here as input quiescent currents is $I_{IO} = 250$ nA within the operating range of the input voltage. The external wiring of the analog signal switch should therefore always make provision for a direct current path in the input circuit across which the input quiescent current can flow.

If the input quiescent currents are applied to the inputs via resistors R_V, the voltage drops which occur across them will affect the input dc voltages U_I. To ensure that the effective input dc voltage U_I and hence also the output dc voltage U_Q changes as little as possible when another input is switched to, it is advisable to choose one value for all the series resistors R_V connected to the inputs of a channel. When the analog signal switches are used in stereo amplifiers, similar series resistors will also be used, in addition, in parallel switch stages to ensure that the voltage drops across the series resistors affect both channels equally.

The current I_I occurring actually at an input may deviate from the average current I_{IO} applying to the whole switch by a value I_F which in the common terminology for operational amplifiers is called the input offset current:

$$I_I = I_{IO} + I_F$$

The offset current I_F can be positive or negative. Its maximum value $I_{Fmax} = 200$ nA and the rms value over many samples amounts to only $I_{Ftyp.} = 20$ nA. These values apply both to the amplifier inputs of a channel and to associated inputs of several channels in an analog signal switch.

Even if the series resistors R_V are equal and at zero signal voltage the input dc voltage and hence also the output dc voltage may still change when the circuit is switched over, because of the different input offset currents in the individual inputs. There is another cause for this effect, however, i.e., different input offset voltages U_F at the individual amplifier inputs. The input offset voltage is here the dc voltage which must be applied to the input in addition to the bias, so that the output dc voltage assumes the same value as the original bias without external load. This input offset voltage may be equally positive or negative; its maximum value is $U_{F\,max} = 10$ mV and the rms value over many specimens is $U_{Ftype} = 2$ mV.

SENSITIVITY OF THE INPUT OFFSET VOLTAGE AND THE OFFSET INPUT CURRENT TO FLUCTUATIONS IN SUPPLY VOLTAGES

Both the input offset voltage U_F and the input quiescent current I_R are to some extent dependent on the supply voltage U_p. Fluctuations in the supply dc voltage U_p therefore cause U_f and I_r of the input offset voltage and the input quiescent current to fluctuate as well, which, in turn, produce fluctuations in the output voltage U_q. This effect of the supply voltage on the output voltage can be described quantitatively by the quantities, normal for operational amplifiers, of the sensitivity H_{UF} of the input offset voltage and the sensitivity H_{IR} of the input quiescent current. For the fluctuations of the output voltage caused by supply voltage fluctuations it applies in the case of decaying signal input voltage that:

$$U_q = HV_u\,U_p.$$

$$\text{with } H = H_{UF} + \frac{Z_G}{2}\,H_{IR}$$

where Z_G = rms generator impedance in the input circuit at the frequency of the supply voltage fluctuations. At low frequencies the quantities H_{UF} and H_{IR} are real and independent of frequency. Typical values for the analog signal switches TDA 1028 and TDA 1029 are approximately $H_{UF} = 5 \times 10^{-5}$ and $H_{IR} = 5 \times 10^{-9}$ A/V.

INHERENT NOISE

The inherent noise of the circuits can be described by two noise sources at the input—a voltage source U_{ir} and a current course I_{ir}—the noise voltage source being in series, and the noise current source in parallel with the input and the subsequent circuit is then regarded as noise-free. For the output noise voltage U_{qr} ultimately of interest and assuming that the external input noise signal only arises from the thermal noise of the generator resistance R'_G, it then applies that:

$$U_{qr} = V_u \sqrt{U_{ir}^2 + R_G^{12} I_{ir}^2 + U_r R_G^2}$$

Where U_{ir} = equivalent input noise voltage, I_{ir} = equivalent input noise current and U_{rRG} = thermal noise voltage caused by the rms generator resistance R^1_G in the input circuit, which is equal to:

$$U_{rRG} = \sqrt{4kT \, R'_G \, b}$$

Whereas the thermal noise voltage depends only by the width b of the transmitted band extending from f_1 to f_2 (white noise), the quantities U_{ir} and I_{ir} are also affected by the position of this frequency band. They may be determined by integration of the spectra of the narrow band noise $(U_{ir}/V f)$ and $(I_{ir}/V f)$ shown plotted in Figs. 21-6 and 21-7 as a function of the transmitted frequency band:

$$\left(\int_{f_1}^{f_2} \left(\frac{U_{ir}}{\sqrt{f}} \right) df \right)^{\frac{1}{2}}$$

The rise of the noise spectra at decreasing frequency is caused by the effect of the excess noise (1/f noise) in the semiconductor elements. In the TDA 1028 and TDA 1029 circuits, typical values for the frequency range of 20 Hz to 20 kHz are for example, $U_{ir} = 3.5$ μV for the equivalent input noise voltage and $I_{ir} = 50$ pA for the equivalent input noise current.

CONTROL VOLTAGES AND CONTROL CURRENTS FOR THE SWITCH DRIVE

As already mentioned in the circuit description, the analog

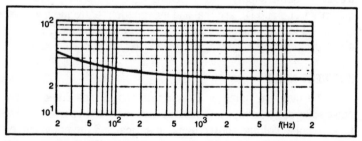

Fig. 21-6. Equivalent input noise voltage referred to 1 kHz bandwidth as a function of frequency.

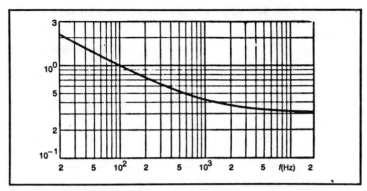

Fig. 21-7. Equivalent input noise current referred to 1 kHz bandwidth as a function of frequency.

signal switch is changed over by switching signals applied externally across an internal switch drive. As seen from Figs. 21-1 and 21-2, the circuit TDA 1028 contains two separate stages for the switch drive, each with an input, for its two independently operating double-pole switches, whereas the TDA 1029 circuit only has one stage with three inputs to drive the switch for its double-pole change-over switch with four switch settings.

As regards the external switch voltages U_S for the switch drive, a distinction must be made between a HIGH-state H with voltages greater than 3.3 V and a LOW-state L with voltages of less than 2.1 V. The input currents at the control inputs are small in the HIGH-state, being in the region of $-0.1\ \mu A < I_{SH} < 1\ \mu A$; in the LOW-state they may assume values as high as $-I_{SL} \leqslant 200\ \mu A$ (the negative sign in front of a current means that the current flows out of a terminal of the integrated circuit). The switches may also be brought to the LOW-state by means of a current by introducing a current $-I_{SL}$ between 200 μ A and 50 mA into the respective control input.

Expressed in relay terminology, a switch will be at HIGH potential at the control input (or the control inputs) in the break position, where the 'break contacts' are closed; at LOW potential at one control input the switch arrives in the associated make position where the appropriate 'make contacts' are activated. The HIGH-state is obtained automatically if the appropriate control input is left open and if it is ensured that the externally acting resistance between the control terminal and earth is greater than 33 M ohm. The simplest way of obtaining the LOW state is by connecting the relevent control input to earth by way of an external contact.

Table 21-1. Switch Drive with Analog Signal Switch TDA 1028.

Switch	Switching voltage U_S	Active signal inputs	Interconnected terminals
S_I, S_{II}	$U_{1/16}$ H $U_{1/16}$ L	I - 1, II - 1 I - 2, II - 2	2 - 4, 15 - 13 3 - 4, 14 - 13
S_{II}, S_{IV}	$U_{8/16}$ H $U_{8/16}$ L	III - 1, IV - 1 III - 2, IV - 2	7 - 5, 10 - 12 6 - 5, 11 - 12

Tables 21-1 and 21-2 show the control voltages U_S at which the indicated analog signal inputs and outputs of the TDA 1028 and TDA 1029 circuits are interconnected. In the TDA 1028 with only one control input per internal switch and two switch settings, the LOW potential at the input sets the switch to one position and the HIGH potential sets it to the other position. The situation is slightly different in the TDA 1029 with three control inputs and four switch settings. Here the particular switch setting is determined by the LOW potential at the input with the lowest terminal number. If LOW potential is not applied to any of the control inputs, the switch will be in setting 1. An internal interlock arrangement in the switch drive prevents several inputs being connected simultaneously to one output. For this reason, the voltages marked X in Table 21-2 may be both LOW and HIGH, provided that the voltages U_{SL} marked L are less than 1.5 V.

SIGNAL SOURCE SWITCH

The analog switches TDA 1028 and TDA 1029 are particularly suitable for use as signal source switches to be operated electrically which is exactly the field of application for which the circuits were especially developed. With the TDA 1028 circuit switching can take

Table 21-2. Switch Drive with Analog Signal Switch TDA 1029.

Switching voltages U_S			Active signal inputs	Interconnected terminals
$U_{11/16}$	$U_{12/16}$	$U_{13/16}$		
H	H	H	I - 1, II - 1	1 - 15, 5 - 9
H	H	L	I - 2, II - 2	2 - 15, 6 - 9
H	L	X	I - 3, II - 3	3 - 15, 7 - 9
L	X	X	I - 4, II - 4	4 - 15, 8 - 9

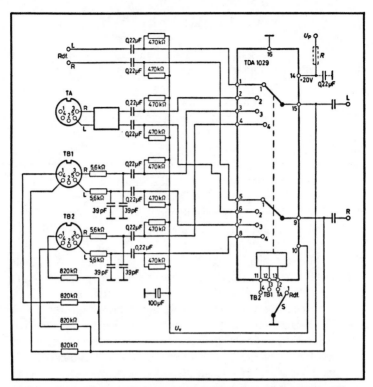

Fig. 21-8. Practical circuit for a signal source switch for four signal sources (e.g., radio Rdf, pickup TA, recorders TB1 and TB2).

place between two signal sources and the TDA 1029 circuit is capable of switching between four signal sources. This number can be further increased by connecting the analog signal switches in cascade. The af section of radio sets in the middle and high price ranges must normally be capable of being connected (at least) to the radio receiving section (Rdf.), to a record player (pickup TA) and to one or two tape recorders (TB1, TB2).

A circuit in which the switching to the above-mentioned four different signal sources takes place with an analog signal switch TDA 1029 and which corresponds largely to the theoretical circuit diagram of Fig. 21-4 is shown in Fig. 21-8. The signal source switch is designed for two channels, normally the left and right stereo channels. The actual switching is effected by the four-stage mechanical switch S connected at one end to earth, through which no af signals pass and which can, therefore, be inserted without problem in any place in the equipment. The signal sources are all

183

coupled capacitively to the switch and the bias U_v is applied to the inputs via 470 ohm resistors from the internal bias source.

The af intelligence signals are applied to the inputs of the signal source switch through leads of shorter or longer length in which rf signals can readily penetrate. Interference due to penetrated *rf* signals can be simply avoided, however, by means of suitable single RC networks connected immediately prior to the signal inputs of the analog signal switches. The RC network must attenuate the rf signals sufficiently without, however, noticeably affecting the af intelligence signals. The resistance of the RC network by which the input current is suitably limited at the same time to protect the circuit should have a value which ensures that the active generator impedance, and hence the crosstalk, does not increase to the extent where it causes interference. Finally the capacitance in the RC network should be sufficiently small to avoid the input impedance of the circuit decreasing noticeably in the upper af range. A suitable compromise must be sought between these partly contradictory requirements in the dimensioning of the RC networks.

In the circuit shown in Fig. 21-8 suitable low-pass filters (5.6 k ohm, 39 pF) are connected to the tape recorder inputs TB1 and TB2. The leads for the radio af signals need not be provided with special *rf* low-pass filters, as the output resistance of the stereo decoder is sufficiently low and rf signals are attenuated before the record player inputs, if such low-pass filters are already contained in the phono equalizing network.

For tape recording, 820 k ohm resistors are inserted between the outputs of the signal source switch and the terminals 1 and 4 of the tape recorder sockets. The resistance value is obtained from the af voltage at the circuit outputs which amounts at a degree of modulation m = 1 to about $U_{a\,rms}$ = 1V, and from the dynamic range for 100 percent modulation at the tape recorder inputs, amounting to between 0.1 m V/kΩ and 2 m V/kΩ. If a value of 0.8 m V/kΩ is chosen for the tape recorder inputs to obtain adequate separation for reasons of tolerance particularly from the upper limiting value, the resistance value will be exactly that indicated. The cases in which coupling-out capacitances are required at the outputs, will be discussed later.

In the simple circuit shown in Fig. 21-8, the crosstalk from a non-switched in signal source to the switched through input and hence also to the output of this channel cannot under certain circumstances be low enough. The crosstalk is governed by the generator impedance of the input to which crosstalk takes place and by the mechanical construction of the circuit, particularly the input sockets which are used. For these reasons it is impossible to give general data on crosstalk.

In general, however, the crosstalk attenuation increases with decreasing generator impedance at the input of the channel to which crosstalk takes place. As far as crosstalk is concerned, it would therefore be particularly suitable by means of input emitter followers to reduce the generator impedances active for the integrated analog signal switch considerably. Under otherwise equal conditions the worst possible values obtained by these measures for the crosstalk attenuation are a U1 = approximately 80 dB at f = 10 kHz and approximately 74 dB at f = 20 kHz. An input circuit with emitter followers for one channel is shown in Fig. 21-9. Such a circuit is required only in the channels with high generator impedance (for the tape recorder inputs) but can be omitted for the record player input with equilization amplifier and for the radio input.

To reduce crosstalk to a minimum, the emitter followers should be arranged as near as possible to the input sockets. Because of the very low emitter follower output resistance, the wiring between the emitter follower output and the analog signal input is

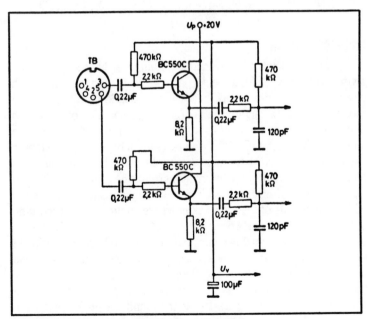

Fig. 21-9. Example of input circuit with emitter follower.

not critical and can be designed with very little restriction. To reduce noise, the emitter direct current of the stages must be as small as possible, yet should be large to keep distortions to a minimum, however. With the low-noise transistor type BC 550 C shown in Fig. 21-9, a satisfactory compromise has been obtained with a current of $E_E = 1.3$ mA.

As illustrated by the following comparison of several properties of the circuits with and without emitter followers, the use of additional emitter followers, even with suitable dimensioning and disregarding additional cost, is not always advantageous.

The following shows the total distortion (k_{ges}) of a signal source switch using the TDA 1029 is given at an output voltage $U_{a\,rms} = 5$ V and a load resistance $R_L = 47$ k at two frequencies, with and without emitter followers.

Total distortion k_{ges} at $U_{a\,rms} = 5$

Frequency f	without emitter follower (Fig. 21-8)	with emitter follower (Fig. 21-9)
1 kHz	0.02 percent	0.10 percent
20 kHz	0.03 percent	0.11 percent

MONITOR SWITCH

A monitor switch is in principle nothing else but a signal source switch which is used for a special purpose. With tape recorders with separate recording-playback heads it is possible with the aid of a monitor switch to carry out a pre/post tape comparison, to which end, according to the setting of the monitor switch, either the signals of the source selected with the switch or the signals coming from the playback head must be applied to the subsequent af amplifier. The monitor switch must therefore be connected after the signal source switch and a special monitor input must be provided to which the signals of the playback head are applied in the recording setting of the tape recorder.

One half of an analog signal switch TDA 1028 is very suitable for use as an electronically switchable monitor switch, only one double-pole switch with two settings being required to change over the monitor in two channels. The theoretical circuit diagram of such a switch is shown in Fig. 21-10. As the monitor signals come from the external tape recorder, a special monitor socket must be fitted

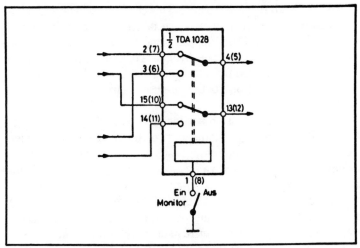

Fig. 21-10. Monitor switch using one half of TDA 1028.

in the af section of the radio and control equipment. In addition, coupling capacitors and rf low-pass filters like those used in the input circuits of the signal source switch, terminal 10 of the TDA 1029, applying it via resistors of, say, 470 ohms to the signal inputs of the monitor switch. To reduce switching clicks to an acceptable level, the signal source switch should also be coupled capacitively to the monitor switch. To ensure that crosstalk from the output circuit of the signal source switch to the input of the monitor switch is reduced to a minimum when the monitor is in the ON position, emitter followers should be inserted in the input circuit for the monitor signal.

MONO-STEREO SWITCH

The requirement to be complied with by a mono-stereo switch is that in the 'Stereo' setting the signals of both channels pass through the switch largely unaffected and that in the 'Mono' setting the signals of each individual channel are passed to both outputs with a transmission ratio 0.5. In this way it is possible in the mono setting to produce both a mono reproduction of stereo signals and a reproduction of mono signals present in only one channel through the speakers of both channels. Such a mono-stereo switch employing electronic switching can be obtained with one half of an analog signal switch TDA 1028, as shown in Fig. 21-11.

The generator resistances in the input sources in this circuit must have exactly the same value as the output resistance of the

TDA 1028 circuit (r_o = 400 ohms). This condition will be automatically fulfilled if the input signals are taken from an analog signal switch TDA 1028 or TDA 1029, as is mostly done in practice. The mono-stereo switch is operated by means of a single-pole mechanical switch which is free from af signals and by which in the stereo setting the terminal 1 or 8 of the integrated circuit is connected to earth.

The required circuit properties are obtained by two voltage-dividers R_{T1}, R_{T2} each consisting of two 10 k ohms resistors which are connected between one output each of the integrated circuit and a signal source. The output voltages of the mono-stereo switch are taken from the taps of the two voltage dividers R_{T1}, R_{T2}. The output resistance of the circuit at 10.4 k ohms is thus relatively high, so that it would be helpful, although not absolutely necessary, if the successive stage could have an input resistance $R_I \gg 5$ k ohm.

Figure 21-11 shows that with the switch set to 'Stereo' each output receives only the signal of one input source across the internal amplifier and the divider resistor R_{T1} as well as direct across the other divider resistor R_{T2}. The signals of both channels are thus independent of each other and with no load applied the voltage at the output will be the same as that at the input. In the 'Mono' setting, each input signal arrives at both outputs; however,

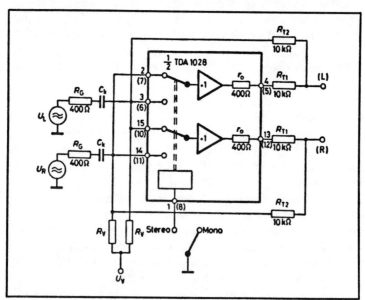

Fig. 21-11. Mono-stereo switch using one half of TDA 1028.

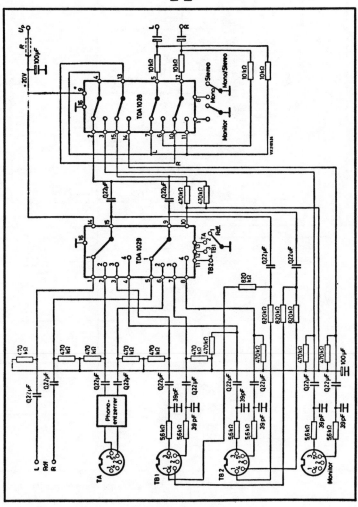

Fig. 21-12. Input section of an *af* preamplifier.

189

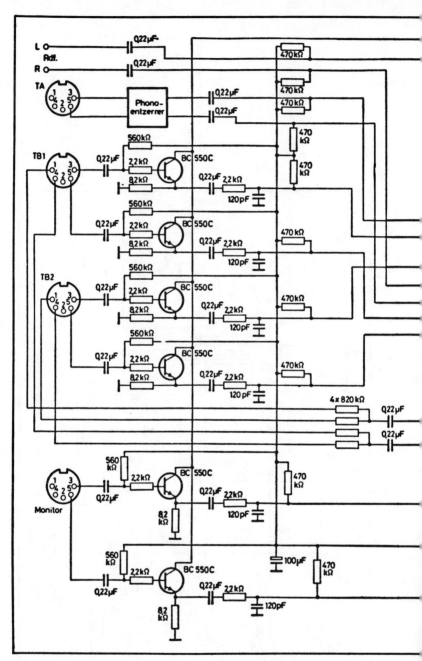

Fig. 21-13. Input section as Fig. 21-12 but with emitter followers inserted for recorder and monitor signals.

190

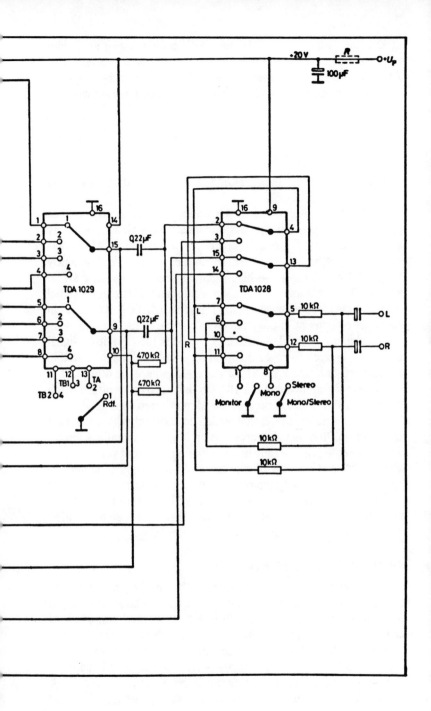

191

at the one output across the internal amplifier and the divider resistor R_{T1} and at the other, across the second divider resistor R_{T2}. As the active generator resistances in these two branches referred to the output are equal in value, the voltage at each output when no load is applied:

$$U_a = \frac{1}{2}(U_R + U_L)$$

Both input signals are thus related to each other additively equally weighted, as required.

If the Mono-Stereo switch is to operate independently of other parts of the circuit, the input signals must be coupled capacitively and a suitable bias U_V must be applied across sufficiently high bias resistors R_V of, say, 470 k ohms. The bias can, of course, frequently be taken from the internal bias supply source of an analog signal switch TDA 1029 used in the same equipment. If the input signal source is also formed by an analog signal switch TDA 1029 used in the same equipment. If the input signal source is also formed by an analog signal switch TDA 1028 or TDA 1029, the capacitive signal coupling can often be replaced by direct coupling. Examples in this respect are shown in Figs. 21-12 and 21-13, where the other half of the analog signal switch TDA 1028 to that used as monitor switch represents the signal source of the Mono-Stereo switch.

Of the data on the circuit which considerably differs in part from those of the signal source switch (e.g., the monitor switch), should be noted the already mentioned higher output resistance of about 5 kΩ and the low output noise. At the required generator resistance $R_G = 400\Omega$ a typical noise voltage in the frequency range of 20 Hz to 20 kHz at the output of the Mono-Stereo switch will amount to 2.2 μV, which is only half that of the signal source switches with low generator resistance described.

SWITCHABLE ACTIVE RC FILTER

The switched through signal paths of the analog signal switches TDA 1028 and TDA 1029 contain amplifiers, making it possible to build active RC filters with these circuits. It is, of course, necessary to use filter circuits which are designed for a voltage gain $V_u = 1$ of the filter amplifier, as the voltage gain of the interconnected channels is exactly unity when no load is applied. The principal advantage of using the circuits TDA 1028 and TDA 1029 in active RC-filters is that it is possible without additional expense and by means of *dc* voltage signals to change over between filter circuits with different frequency characteristics and naturally also to switch to a frequency-independent gain ('filter off'). High- and low-pass filters are very important for the af section of radio equipment. The high-pass filters are used, for example, as rumble filters to attenuate the rumble signals coming from the record player to an acceptable level. They are also suitable as a 'subsonic filter', however, to dampen interference signals below the audio spectrum which at high amplitudes modulate the output stage and the loudspeaker to an unnecessarily great and undesirable extent. Noise can be very disturbing when old records are played back or on stereo reception of stations not having a particularly high field-strength. By means of a suitable low-pass filter, here appropriately called a 'noise filter', the high frequencies can be cut off and the reproduction improved.

It is not easy in this respect to give the optimum frequency characteristic. It certainly depends on the spectrum of the noise and the intelligence signal; for this reason alone it is already impossible to give a definite optimum frequency characteristic. In the applications discussed we may say with certainty as follows, however: If the intelligence signal is to be affected as little as possible by the filter, it may virtually not change in the most important part of the af range, hence from 120 Hz to 6 kHz (this means a quality of roughly 80 percent in the case of music compared with 100 percent at a frequency range not cut down to any extent). The filter curves should thus vary gradually in its response range. Furthermore, they should drop sufficiently steeply beyond the limiting frequency f_0 to suppress the noise signals in the stop band adequately, this band being less important to the intelligence signal. More important than a very high distant selectivity is that the transition from the gradual variation in the transmission band to the drop in the stop band takes place as quickly as possible without the frequency response rising visibly in the region of the cut-off frequency.

The greater the frequency response in the stop band, the quicker the transition from the transmission band to the steep drop will take place and the greater the visible rise in frequency response, the greater the overshoot of the output voltage will be when a step function signal is applied. Strong overshoot in response to the step function is certainly undesirable and filters with very steep curves with a quick transition from transmission to stop band, apart from being very expensive, are therefore unsuitable. A relative overshoot of less than about 10 percent might be acceptable, particularly if it is borne in mind that signals with very steep pulse edges do not occur in the *af* channel.

It is clear from these considerations which type of filter will be most suitable for the application under review. Although Chebishev filters have steeper flanks compared with other types of filter, their frequency pass-band contains very small peaks and valleys, so that they are unsuitable for the case in question. Bessel filters have virtually no overshoot when step function signals are applied, but the transition from the pass-band in the stop band extends over a relatively large frequency range. Butterworth filters appear to be most suitable for the case in question; they have no ripple or show any rise in the frequency response of the pass-band and cause only slight overshoot when step function signals are applied. As the overshoot in response to step functions is less than 10 percent when a Butterworth filter of the third order is used, the drop in the stop-band of 18 dB/octave is satisfactory and the cost is relatively low (see Fig. 21-14). The cut-off frequency of the filter should be so chosen that a good compromise is obtained between maximum attenuation of the disturbing signal and minimum impairment of the intelligence signal.

In addition to the frequency characteristic, other properties of the filter must be taken into account in the dimensioning, e.g., the

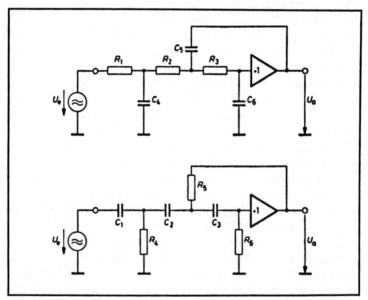

Fig. 21-14. Low-pass (top) and high-pass (bottom) filters of the third order.

inherent noise of the circuit. If a filter is used for the attenuation of noise signals, it should obviously contribute as little as possible to the noise itself. In active RC-filter networks it may well occur that by the existing feedback the signals of the internal noise sources appear amplified to a greater or lesser extent at the output in certain frequency ranges, even to a much greater extent than the input signal. This is possible because the internal noise signals are active at different places of the feedback circuit than the input signal. It is thus very important to dimension the filter in such a way that the inherent noise has as little effect as possible.

The internal noise at the output is governed, in addition to the relationships between the components with respect to each other, also by the frequency response of the filter. The quicker the transition from pass-band to stop-band the greater the transgression in the amplitude frequency curve, and the more steeply the filter curve drops in the stop-band, the stronger the inherent noise signals will generally be boosted. To obtain a sufficiently low level of inherent noise, it may consequently be necessary to compromise with regard to frequency response and to accept a slightly broader transition from pass-band to stop-band.

The inherent noise at the filter output may also be reduced by designing the RC-filter network so that the dc resistance is as low as possible. The effect of the equivalent input noise current source of the amplifier will then be reduced. Limits are imposed here for practical reasons, however, as the filter may not present too great a load to the input signal source and also, via the feedback, to its own output. It should be pointed out here that with switchable filters the RC network in the disconnected filter (where it may well be a question of a signal path with transmission factor not affected by frequency) may only have very little effect on the signal path of the connected filter. Moreover, the filter circuit should be so dimensioned that the sensitivity of the filter properties to changes in value of the components is very low. An important practical aspect is finally that only standard values (possibly of the E6 or E12 series) should be used for components in filters to reduce the costs and simplify storage. This requirement can naturally be fulfilled only when certain allowances are made as to the frequency response. Efforts can always be made to find a dimensioning with standard values at which the deviations from the required frequency response are as little as possible.

In accordance with the above, the requirements imposed on filters are partly contradictory and the best possible compromise should thus be sought for the dimensions, always bearing in mind which filter properties are regarded as being most important.

As the voltage gain of the analog signal switch used here for the filter design (no-load) is unity, the modulation problems arising with the filter circuits proposed later are no more critical than those encountered in the circuits already discussed above, so that questions of modulation with respect to the filters needs no further discussion, but see also Chapter 25 on filters.

INTERSTAGE COUPLING

At the external inputs of an *af* preamplifier containing an analog signal switch TDA 1029 (or TDA 1028) in the input stage, the best coupling to use is capacitive signal coupling as this provides electrical isolation and enables the required bias U_v to be applied readily to the inputs of the integrated circuit. For similar reasons capacitive coupling could also be used generally at the output of the preamplifier.

If two stages containing the analog switches TDA 1028 or TDA 1029 are connected in cascade, it is often possible to employ resistive/inductive coupling, i.e., a direct connection between the preceding stage outputs and the inputs of the succeeding stage. With the analog signal switches with a voltage gain $V_u = 1$ the input and output voltages are, in fact, equal, disregarding the input offset voltage. The addition of the input offset voltages of the two stages does not generally give rise to any disturbance.

There is, of course, a circuit configuration where under certain circumstances, even when two stages with the analog signal switches discussed here are connected in cascade, capacitive coupling is preferred to resistive/inductive, i.e., direct coupling (in the case of the monitor switch where in the 'off' and 'on' settings respectively the outputs of the previous signal source switch and the monitor socket are connected to the interconnected inputs). The stages are thus, in fact, connected in cascade only in the 'off' setting.

Chapter 22

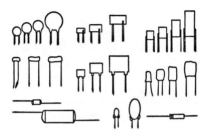

Electromagnets

A basic formula for the pull of an electromagnet is:

$$\text{Pull} = \frac{B^2 A}{4464 \times 10^4} \text{pounds}$$

where B is the flux density of the core (lines per square inch) and A is the cross sectional area of the core in square inches.

If soft iron is used for the core, and the cell is designed to fully saturate it, then B can be taken as 80,000 lines per square inch. Typical leakage losses can be 20 - 30 percent, and iron losses of the order of 10 percent. So, for a practical working formula take a value of 50,000 lines per square inch for B, when:

$$\text{Pull} = \frac{(90,000)^2}{4,464 \times 10} \times A$$

$$= 56 \times A \text{ pounds}$$

Example: Estimate the pull of an electromagnet (with a coil designed to fully saturate a soft iron core) having a core diameter of 1 inch.

$$\text{Ans. core area (A)} = .7854 \times 1^2$$
$$= .7854 \text{ sq. inch}$$
$$\text{Hence Pull} = 56 \times .7854$$
$$= 44 \text{ lb. approx.}$$

More usually the problem is to design a suitable coil to provide saturation of the core. This can be done using the formula which relates flux density to coil ampere turns, viz:

$$\text{Ampere turns (IT)} = \frac{0.8 \text{ BL}}{\mu}$$

where L is the length of core
B is the permeability of the core material

For soft iron (approximately):

$$\text{IT} = 20 \text{ per core}$$
$$\mu = 50 \text{ per inch}$$

DESIGN OF COMPLETE NEGATIVE CIRCUIT

The effective (negative) resistance in a complete magnetic circuit (Fig. 22-1) is known as the *reluctance*, given by:

$$S = \frac{\text{length of path(L)}}{\mu \times A}$$

where A is the cross section area.

Example calculation for an iron circuit 2 cm square with L1 = 10 cm and L2 = 20 cm, permeability of part 1 taken as 1,000 and part B 1,250 (typical working figures for soft iron). Note that the path lengths are taken as centerline lengths.

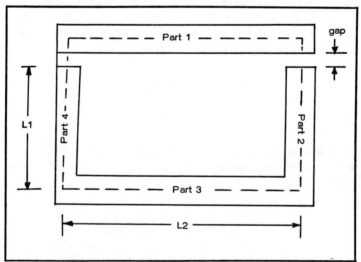

Fig. 22-1. Magnetic circuit geometry for electromagnet design.

$$\text{Reluctance of part 1} = \frac{(20-2)}{1,000 \times 4} = 0.0045 \text{ units}$$

$$\text{Reluctance of part 2} = \frac{(10-1)}{1,250 \times 4} = 0.0018 \text{ units}$$

$$\text{Reluctance of part 3} = \frac{(10-1)}{1,250 \times 4} = 0.0018 \text{ units}$$

$$\text{Reluctance of part 4} = \frac{(20-2)}{1,250 \times 4} = 0.0036 \text{ units}$$

$$\text{Reluctance of air gaps} = \frac{(2 \times 0.2)}{1 \times 4} = 0.1 \text{ units}$$

(Note: permeability of air is 1.0)

$$\text{Total reluctance} = 0.1117 \text{ units}$$

In a magnetic circuit:

Magnetomotive Force (MMF) = Reluctance × Flux

(Note: this is similar to current flow in an electric circuit where current = flux, resistance = reluctance, and voltage = MMF. The MMF is supplied by the current flowing through the coils and is given by:

$$\text{MMF} = 0.4\pi IT$$

where I is the current through the coil
and T is the number of turns

Taking as an example a coil of 1,000 turns carrying a current of 0.75 amps:

$$\text{MMF} = 0.4\pi \times (2 \times 1,000) \times 0.75$$
$$= 1884 \text{ units}$$

$$\text{Total Flux} = \frac{\text{MMF}}{\text{Reluctance}}$$

$$= \frac{1884}{0.1117}$$

$$16,866 \text{ lines}$$

$$\text{Thus flux density} = \frac{\text{total flux}}{A}$$

$$= \frac{16,866}{4}$$

$$= 4216 \text{ lines per sq. cm}$$

FULLY SATURATED CIRCUIT

To *saturate* a soft iron core in such a circuit (for maximum power), flux density required would be 12,400 lines per sq.cm (80,000 lines per sq.inch). This would require a coil with a higher ampere turns figure:

Flux density required = 12,400 lines per sq. inch
Total flux required = 4 × 12,400
= 49,600 lines

Thus (for the same reluctance as previously calculated):

MMF required = 49,600 × 0.1117
= 5540 units
Now MMF = 0.4 π IT

or IT (ampere turns) required = 4,410 units

This then establishes the *coil performance* required. Thus if 1,000 turns are retained in each coil, current flow through the coil must be 4,410/2,000 = 2.2 amps (approx.).

Note: ampere turns (IT) are dependent on the product of current (I) and number of turns (T). For a required value of ampere turns, the number of turns used can be increased to *decrease* current, and vice versa. Thus if in this example it is desired to limit the circuit to 1.2 amp, say, the number of turns required on the coils must be increased.

For IT = 4,410 at I = 1.2

Number of turns required $\dfrac{4410}{1.2}$

= 3,675

or 1,840 turns on each coil, say.

ALLOWANCE FOR LOSSES

Allow for losses by magnetic leakage in a circuit of this type, especially if it includes an air gap. The loss factor may be anything from 1.1 to 1.5, depending on materials, design and size of air gaps. Allow, say, 1.25 in the example above. Thus: number of turns required on each coil to allow for likely magnetic leakage: (1,840 × 1.25 = 2,300 turns).

SINGLE COIL MAGNETIC CIRCUITS

A single coil magnetic circuit, is relatively inefficient as an electromagnet (Fig. 22-2A) because there is no 'return path', except

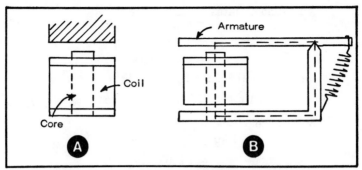

Fig. 22-2. Magnetic circuit with return path (right) is more efficient.

through air. The twin pole (horseshoe type) electromagnet of Fig. 22-1 is thus a better design for achieving maximum pull from a given ampere turn.

A single coil electromagnet immediately becomes more efficient if it is provided with an iron return path as in Fig. 22-2B. This is typical of relay design, incorporating a pivoted armature. Coil design is then worked out on the lines previously described:

(i) Determine the total reluctance of the circuit, including the air gap.

(ii) Work out the total flux required (flux density multiplied by coil core crosssectional area).

(iii) Calculate the MMF required, viz:

$$\text{MMF} = \text{total flux} \times \text{reluctance}$$

(iv) From this, work out the ampere turns required:

$$\text{ampere turns (IT)} = \frac{\text{MMF}}{0.4\,\pi} = \frac{\text{MMF}}{1.257}$$

SOLENOID DESIGN

A solenoid is a coil wound on a coil of non-magnetic material and thus effectively an air-cored coil. The field strength (H) generated inside the solenoid is given by:

$$H = 1.257 \text{ ampere turns per cm}$$
$$= 3.2 \text{ ampere turns per inch}$$

In practical application, it is the *magnetizing force* of the solenoid.

Example: Calculate the field strength generated inside a solenoid consisting of a 1,200 turn coil wound to a length of 2 inches. The circuit flowing through the coil is 0.8 amps.

Ans: Ampere turns = 1,200 × 0.8
= 960
Field strength = 3.2 × 960 × 2
= 6,144 units

The magnetomotive force generated by such a solenoid is:

MMF = 1.257 × ampere turns

Using the example above:

MMF = 1.257 × 960
= 1,206 units

If L is the length of the solenoid and D its diameter, reluctance

of solenoid $= \dfrac{L}{.78540^2}$

Note: This ignores the reluctance of the return path through air, but this is justified in this case because of the very much greater 'sectional area' of the air embraced by the return path. In other words, it is fair enough to assume that the whole of the MMF is utilized in driving the flux through the reluctance $L /.78540^2$.

Dc Solenoids

For simplified design the pull of a *dc* solenoid can be calculated directly from the following formula:

P = k IT A/L

where A   is the cross sectional area of the plunger
L   is the length of the coil
k   is a constant depending both on the units employed and the relative proportions and positions of solenoid coil and plunger.

Maximum pull is normally given when the length of plunger entering the coil is between about 30 and 70 percent. Thus $P_{max} = k^1$ IT A/L.

The constant k^1 is now dependent only on the units employed and the relative coil and plunger geometry. For English units with pull (P) in pounds, A in sq. inches and L in inches, k^1 typically ranges between 0.005 and 0.1. For pull in kilograms, A in sq.cm, and L in centimeters, typical values of k^1 range between 0.03 and 0.6. Actual design values are best obtained from empirical tests, or based on the known performance of solenoids of similar geometric proportions.

202

The pull of a dc solenoid can be increased substantially by providing an iron circuit to complete a magnetic 'loop' circuit, and thus increase its efficiency. The pull available then comprises that provided by the solenoid coil (Ps) and also a further pull provided by an iron circuit of suitable design (Pm) terminating in a pole piece in line with the plunger, which also acts as an end stop.

$$P = Ps + Pm$$

The value of Pm likely to be achieved can be estimated on the same lines as for electromagnets, but will vary over the length of the stroke because of the changing air gap. Thus Pm will be a maximum at the end of the inward stroke.

Ac Solenoids

Ac solenoids are normally designed on empirical lines because of the complex relationship of the individual factors involved. For preliminary design the transformer equation may be used to relate reactive voltage in terms of total flux, number of turns and frequency. Written as a solution for number of turns required (T):

$$T = \frac{E}{4.4 \ BAf} \times 10^8$$

where $E =$ reactive voltage (85 percent of the normal voltage is a typical design figure).

$f =$ frequency

Adjustment can then be made on the basis of practical tests:

☐ If the pull is less than that required, then the number of turns in the coil should be increased to reduce the heating.

☐ If the pull is excessive, then the number of turns in the coil should be increased to reduce the heating.

Coil heating must be checked (the pull required must be obtained at an acceptable working temperature for the coil).

Chapter 23

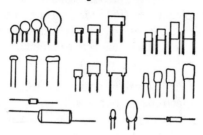

Dc Coil Design

The two primary parameters in coil design are:

☐ The magnetic field strength required, which is normally expressed in *ampere turns*:

ampere turns = I.T. where I = current in amps
T = number of turns on coil

☐ The rating of the coil, which is a measure of the peak temperature rise of the coil due to resistive losses:

resistive loss = I^2R watts where R = resistance of coil

In practice coil rating is normally expressed in terms of watts per square unit of curved surface area of the coil (watts/sq. inch watts/sq. cm.).

Typical general ratings used for coil designs:

Type of duty	Watts/sq. inch	Watts/sq. cm
continuous	0.25-0.5	0.04-0.08
intermittent, but frequent	1.5-2.5	0.25-0.40
intermittent, occasional (5 minutes in 1 hour)	5.0-7.0	0.80-1.10
intermittent, infrequent (1 minute in 1 hour)	7.0-10.0	1.10-1.55

AMPERE TURNS REQUIRED

Ampere turns required can be calculated from the formula:

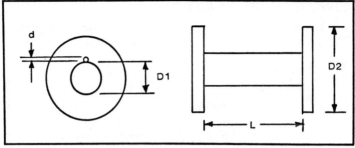

Fig. 23-1. Coil geometry parameters.

$$\text{ampere turns (IT)} = \frac{VT}{R}$$

where V is the voltage across the coil, T is the number of turns, R is the coil resistance.

For a specified ampere turns requirement and a specific coil size of inner and outer diameters D1 and D2 respectively (Fig. 23-1) the approximate bare wire diameter (d) required is given by:

$$d \text{ (inches)} = 1.17 \frac{IT (D1 + D2)}{V} \times 10^{-3}$$

$$\text{or } d = 1.17 \frac{IT (D1 + D1)}{V}$$

where d is in thousandths of an inch (mils). Note: this formula only applies to copper wire.

Example: Calculate the wire size required to give an ampere turns value of 30,000 wound to fill a bobbin where D1 = ¼ inch and D2 = ½ inch. Coil voltage available is 6 volts.

Using the formula:

$$d = 1.17 \sqrt{\frac{30,000 \times (0.25 + 0.5)}{6}}$$

$$= 1.17 \sqrt{3750}$$

$$= 72 \text{ mils}$$

This, in fact is 13 gauge wire. If the calculation had produced a non-standard wire size, it would be necessary to recalculate using a different rating of D2 to adjust as necessary to arrive at an available wire size.

COMPLETE COIL FORMER GEOMETRY

Coil former geometry is covered by the following formulas. The diameter of insulated (covered) wire is designated dc.

Winding depth $= \frac{1}{2}$ (D1 − D2) .... (1)
Winding space $= L/2$ (D1 − D2) .... (2)
No. of turns per layer $= L/dc$ .... (3)

$$\text{No. of layers} = \frac{D1 - D2}{dc} \quad .... (4)$$

$$\text{Length of mean turn} = \frac{\pi}{2} (D1 + D2) .... (5)$$

$$\text{Total number of turns} = \frac{L(D2 - D1)}{2\,dc^2} .... (6)$$

$$\text{Length of wire} = \frac{.7854\ L(D2^2 - D1^2}{dc^2} .... (7)$$

$$\text{Diameter of insulated wire (dc)} = \sqrt{\frac{L(D2 - D1)}{2T}} .... (8)$$

Due allowance must be made for the number of thicknesses involved if paper interlays are used in winding the coil. These formulas are rather easier to use for design calculations as it is possible to start with a selected size of wire and finalize the actual coil geometry to suit. It is still necessary to start with an assumed or 'guesstimated' size of coil, however.

The *length* of wire required is then given by formula (7). The *resistance* of this length of wire of chosen diameter size can then be found from wire tables. The ampere turns of the coil can then be calculated directly from:

$$IT = \frac{VT}{R}$$

Example: Use the same coil former properties as before (D1 = $\frac{1}{4}$ inch and D2 = $\frac{1}{2}$ inch), with a coil length of 1 inch. Selected wire size is 30 gauge enameled, giving a covered wire diameter (dc) of 11.5 mils. Resistance of this size of wire is 105.2 ohms per 1,000 ft (from standard wire tables).

$$\text{Length of wire (from formula 7)} = \frac{.7854(\ (\tfrac{1}{2})^2 - (\tfrac{1}{4})^2)}{11.5^2 \times 10^{-6}}$$

$$= \frac{.7854 \times (3/16)^2}{12.25 \times 10^{-6}}$$

$$= 208.7 \text{ feet}$$

Resistance of this length of wire $= \dfrac{208.7}{1,000} \times 105.2$

Total number of turns (from formula 6) $= \dfrac{1 \times (\frac{1}{2} - \frac{1}{4})}{2 \times (11.5)^2 \times 10^{-6}}$

$$= \frac{0.25}{264.5 \times 10^{-6}}$$

$$= 945$$

If the coil voltage is 6 volts, then:

$$IT = \frac{6 \times 945}{21.955}$$

$$= 258.25$$

If this value is unsatisfactory, then wire size and/or coil dimensions must be altered and formulas recalculated to arrive at a suitable value for ampere turns. Equally, calculation can be started with the ampere turns required and wire diameter to be used, and suitable coil proportions calculated from this. This will involve two 'unknowns', T (the number of turns) and R (wire resistance), where:

$$\frac{T}{R} = \frac{IT}{V}$$

It is easiest to fix T as a definite value, then proceed.

Example: Design requirement is for a 200 ampere turns coil with a 6 volt supply.

Assume number of turns $= 1,000$

Then wire resistance (R) $= \dfrac{T}{IT \times V}$

$$= \frac{1,000}{200 \times 6}$$

$$= 0.83 \text{ ohms}$$

Now, select a possible wire size—say 20 gauge, which has a resistance of 10.35 ohms per 1,000 feet. Length of 20 gauge wire to give 0.83 ohms:

$$= \frac{1,000}{10.35} \times 0.83$$

$$= 80.2 \text{ feet}$$

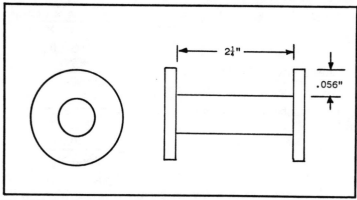

Fig. 23-2. Coil design example sizes.

Check the covered diameter of the 20 gauge wire to be used—e.g., 35 mils with insulation. 'Guesstimate' a suitable coil length—say 2¼ inches. Use formula (3) to determine number of turns per layer:

$$= 2\ 25/0.035$$
$$= 64\ .28$$

Therefore number of layers required to give 1,000 turns is $\frac{1,000}{64} = 15.625 - $ say 16.

Use formula (4) to find winding depth required:

$$(D2 - D1) = 0.035 \times 16$$
$$= 0.56 \text{ inches}$$

Thus a suitable bobbin size works out as shown in Fig. 23-2.

SPACE FACTOR

To allow for possible variations in wiring and actual wire diameters it is usual to allow a 'loss' figure of 5 percent on the actual number of turns a coil former or bobbin will accommodate. The number of turns per inch length of coil former is similarly expressed by 0.95/dc.

There are also two possible ways of layering (Fig. 23-3). Alternating the layering gives maximum *space factor* and will, in fact, provide maximum density of winding, or virtually the full theoretical geometric figure. That is, the 5 percent allowance above can be ignored in practice. Vertical piling of turns will give minimum space factor.

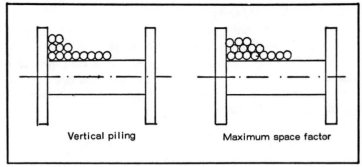

Fig. 23-3. Vertically piled windings take more space.

COIL TEMPERATURE

The *average* temperature of a coil can be determined from 'cold' resistance and 'hot' resistance under operating conditions. A general method which is adequate for most purposes in the case of coils wound from copper wire is to measure the increase in resistance when hot and then assess the coil temperature on the basis of 1°C rise per 0.4 percent increase in resistance:

$$\text{Temperature rise } °C = \frac{250 \ (Rh\text{-}Rt)}{Rc}$$

where Rh = hot resistance
Rc = cold resistance (normal design resistance)

It should be emphasized that the coil temperature given by the above is an *average* figure for the coil. Localized areas of the coil may have lower or higher temperatures.

Example: Coil resistance measured when cold is 42 ohms and when hot is 48 ohms. Calculate the temperature rise in the coil when hot:

$$\text{Temperature rise} = \frac{250 \ (48 - 42)}{42}$$
$$= 35.7°C$$

Table 23-1. Recommended Thicknesses for Paper Interlay.

Wire size (gauge no.) Thickness of paper	16-22 .002 (2 mil)	23-38 .001 (1 mil)	over 38 .0005 (½ mil)

COILS WITH INTERLAYERS

Coils are commonly wound with insulated wire with separate insulation between each layer of the windings. Normally unvarnished paper is used for the interlayers, with special varnished papers, varnished silk, glass fabric, etc., for special duty applications. Recommendations for normal interlayer thicknesses are summarized in Table 23-1.

Chapter 24

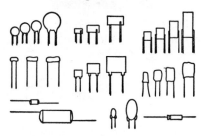

Ac Coil Design

The design of ac solenoids and ac electromagnetic coils is largely derived empirically due to the large number of variables involved, e.g., based on the performance of similar (existing) coils on similar circuits. Where no suitable data is available an experimental prototype coil is usually produced, its performance evaluated under the required working conditions and the winding specification adjusted, as necessary, to arrive at the performance required.

The ohmic resistance of an ac coil is usually low, and thus comprises a small number of turns. The current is determined by the impedance. The limiting factor will then be the temperature rise, which with ac is generated by eddy canals and hysteresis as well as resistive losses.

A practical formula for empirical design is:

$$\text{No. of turns} = \frac{V}{BAf} \times 10^7$$

where V = applied voltage
A = cross sectional area of iron core (sq. inch)
B = flux density of core (lines per sq. inch)
f = frequency

Performance of the coil is evaluated by adjusting the voltage, as necessary, to achieve the desired performance. The correct number of turns for the final design is then calculated as follows:

Number of turns (final design) = Experimental coil turns multiplied by:

$$\frac{\text{design operating volts}}{\begin{array}{c}\text{volts giving desired performance}\\\text{with experimental coil}\end{array}}$$

Example: Flux density required from an ac coil is 80,000 lines/sq. inch where applied voltage is 120 and frequency 60 Hz. Cross section of wire coil is 0.8 sq. inch

$$\text{No. of turns} = \frac{120 \times 10^7}{80,000 \times 0.8 \times 60}$$
$$= 312.5$$

This prototype coil is then checked out by practical testing, adjusting the supply voltage as necessary to achieve the desired result. Suppose these tests show that a coil voltage of 135 is necessary for the prototype to perform satisfactorily. Then number of turns for final design

$$= 312.5 \times \frac{120}{135}$$
$$= 278 \text{ turns.}$$

RF COILS

The inductance of a single layer air-core coil can be calculated with good accuracy from the formula:

$$L \text{ (microhenrys)} = \frac{R^2 N^2}{9R + 101}$$

where R = radius of coil in inches
N = number of turns
I = length of coil in inches

This formula is most useful rewritten as a solution for number of turns required for a given coil diameter and length, viz:

$$N = \sqrt{\frac{L(9R + 101)}{R^2}} = \sqrt{\frac{4L(4.5D + 101)}{D^2}}$$

where D = coil diameter, and L is the inductance value required in microhenrys.

Examples: Inductance required from a single layer air-wired coil is 0.1 microhenry. Projected coil diameter is 0.55 inch and length 1.25 inch. Calculate the number of turns required:

$$N = \sqrt{\frac{4 \times 0.1 \,(4.5 \times 0.55 + 10 \times 1.25)}{(0.55)^2}}$$

$$= \sqrt{\frac{0.4\,(2.475 + 12.5)}{0.3025}}$$

$$= \sqrt{\frac{5.99}{0.3025}}$$

$$= \sqrt{19.80}$$

$$= 4.45$$

Either use 4½ turns or readjust coil proportions (e.g., length or diameter) to arrive at an even number of turns, i.e., 4 or 5).

It will be noted that the actual wire diameter does not affect the inductance. Wire diameter can be selected on the basis of rigidity required, if the coil is unsupported. The wire diameter must, of course, be small enough for the number of turns to be accommodated in the design coil length, with equally spaced turns. Actual wire diameter will, however, modify the diameter. The design of air-core coils should be based on a coil length of at least half the coil diameter, and preferably more. Also the larger the diameter of the coil, the more accurate the formula calculation is likely to be.

It is also obvious that since inductance is affected by coil length, mounting should allow for a permanently fixed length. If necessary, the actual inductance of the coil can be modified by a slight change in length. Any subsequent change in the spacing of the coils—e.g., by mechanical shock—will also affect the inductance value. For this reason air-cored coils are best supported on a rigid former of appropriate diameter. Air-core coils of less than 1 inch diameter are not recommended, unless the likely inductance value can be established by other means than the basic formula calculation.

The inductance of an iron-core coil is given by:

$$L = \frac{A_L^2}{N^2}$$

where A_L is the 'inductance value' of the core material normally expressed in microhenrys. This will be affected both by the core material and the core section.

Basically the presence of an iron core increases the flux density, and hence the inductance, relative to the permeability of the material. Permeability will vary with flux density, up to the point where the core is saturated. Thus the inductance of an iron-core coil is dependent on the current flowing through the coil, up to the point where the core is saturated. With air-core coils the inductance is independent of current since air does not saturate.

Graphic or nongraphic solutions are best applied for the design of iron-core coils related to the specific performance of the material if known; or coils wound to specification(s) provided by the core manufacturers. A certain tolerance is always available since if the coil is wound on a former with an iron-duct core, adjustment of the core provides adjustment of the actual inductance of the coil.

Chapter 25

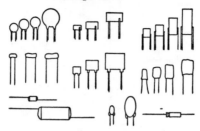

Simple Filters

Two components only are needed to produce a basic filter circuit—a capacitor C and a resistor R. These form a simple network with a *time constant* determined by the values of C and R. This time constant determines the *cut-off* frequency of the filter, given mathematically by:

$$f = \frac{1}{2\pi\ RC}$$

BASIC LOW-PASS FILTER (FIG. 25-1)

With R in series and C in shunt configuration, i.e., across the output, frequencies *lower* than the cut-off frequency of the RC combination are passed without attenuation, higher frequencies being attenuated.

BASIC HIGH-PASS FILTER (FIG. 25-2)

With C in series and R in shunt configuration, frequencies *above* the cut-off frequency are passed without attenuation, and all lower frequencies are attenuated.

BASIC BAND-PASS FILTER (FIG. 25-3)

Band pass filters or bandwidth filters can be produced by combining a low-pass filter in series with a high-pass filter. If the band width is from f_L to f_H, then the cut-off frequency for the

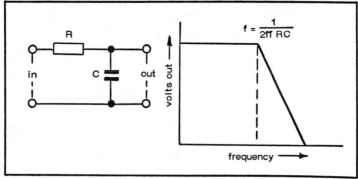

Fig. 25-1. Basic low-pass filter.

low-pass filter is made f_H and that of the high-pass filter f_L. This filter combination will pass frequencies from f_L to f_H (in the desired band).

BASIC BAND-REJECT FILTER (FIG. 25-4)

To produce a band-reject filter, a low-pass filter is used in parallel with a high-pass filter, as in the second diagram. This combination will reject all frequencies within the band f_L to f_H.

The amount of attenuation provided by a filter is expressed by the ratio volts out/volts in, or voltage ratio. This is quoted in decibels (dB)—a 3dB drop being equivalent to a voltage ratio drop from 1.0 to 0.707, or a *power* loss of 50 percent.

Filters may be implemented in various hardware forms (i.e., complete filter circuits), crystal filters and ceramic filters. They are normally 'peak' filters with a specified center frequency and bandwidth. Ceramic filters are designed primarily for use in FM

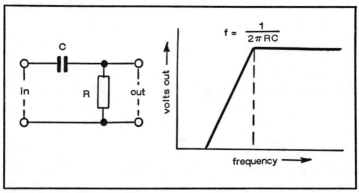

Fig. 25-2. Basic high-pass filter.

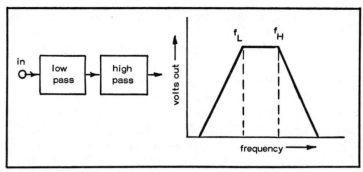

Fig. 25-3. Basic band-pass filter.

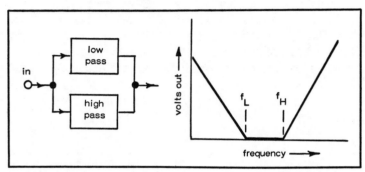

Fig. 25-4. Basic band-reject filter.

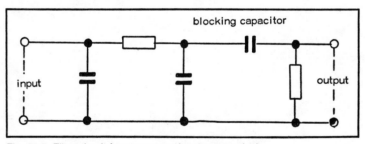

Fig. 25-5. Filter circuit for power supply to remove ripple.

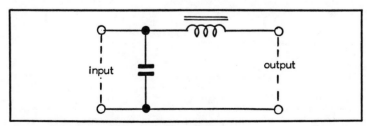

Fig. 25-6. Choke-input filter is a preferred type.

receivers to remove i-f (i.e., have center frequencies corresponding to the i-f). They combine good i-f tuning with good temperature stability and low distortion in a very small package size.

FILTERS FOR POWER SUPPLIES

Ripple remaining in a rectified power supply is at mains frequency with half-wave rectification and twice mains frequency with full-wave rectification. A simple filter circuit (Fig. 25-5) can provide smoothing, but at these frequencies the reactance of C must be small compared with the resistance R. At the same time the value of R must be low in order to avoid unwanted voltage drop and power loss. This means using high values for C—at least 1 μF.

For these reasons a choke-input filter may be preferred (Fig. 25-6). The choke needs to be of special design (swinging-choke) so that its inductance increases substantially as the current falls with the aim of maintaining a substantially constant voltage across it.

Chapter 26

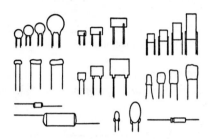

Resonant Circuits

When an inductor (L) and a capacitor (C) are connected in series and fed with an *ac* signal, the reactance of both L and C is frequency dependent, but in opposite ways, viz:

$$\text{Inductive reactance } X_L = 2\,\pi\,fL$$

$$\text{Capacitive reactance } X_C = \frac{1}{2\,\pi\,fC}$$

The practical circuit will also contain some resistance R, in series with L and C (see Fig. 26-1). If the *ac* supply is adjusted to a *low* frequency, the capacitive reactance will be very much larger than R, and the inductive reactance will be much lower than R (and thus also very much lower than the capacitive reactance).

If the *ac* supply is adjusted to a *high* frequency the opposite conditions will apply—inductive reactance much larger than R, and capacitance reactance lower than R and L. Somewhere between these two extremes there will be an *ac* frequency at which the reactances of the capacitance and inductance will be equal, and this is the really interesting point. When inductive reactance (X_L) equals capacitive reactance (X_C), the voltage drops across these two components will be equal but *180 degrees out of phase.* This means the two voltage drops will cancel each other out, with the result that only R is effective as total resistance to current flow. In other words, maximum current will flow through the circuit determined only by the value of R and the applied *ac* voltage.

Working under these conditions the circuit is said to be *reson-*

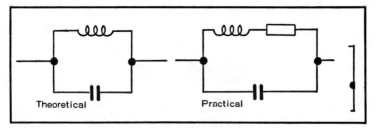

Fig. 26-1. Basic components forming a tuned circuit.

ant. Obviously resonance will occur only at a specific frequency, which is thus called the *resonant frequency*. Its value is given by the simple formula:

$$f = \frac{1}{2 \pi \sqrt{LC}}$$

where f = resonant frequency in Hz
    L = inductance in henrys
    C = capacitance in farads

A more convenient formula to use is:

$$f = \frac{10^6}{2 \pi \sqrt{LC}}$$

where f = resonant frequency in kilohertz (kHz)
    L = inductance in microhenrys (μH)
    C = capacitance in pico farads (pF)

Note that the formula for resonant frequency is not affected by any resistance (R) in the circuit. The presence of resistance does, however, affect the *Quality factor* or *Q* of the circuit. This is a measure of how sharply the circuit can be tuned to resonance, the higher the value of Q the better, in this respect. The actual value of Q is given by:

$$Q = \frac{X}{R}$$

where X is the reactance in ohms of *either* the inductance *or* capacitance at the resonant frequency (they are both the same, so it does not matter which one is taken) and R is the value of the series resistance in ohms.

A surprising feature of this circuit is the very high *magnification* it can produce. For example, suppose the signal strength is 1 mV. The current flowing through the circuit at resonance is then

220

.001/1 = 1 mA (the 'resistance' offered by L is cancelled out by the opposite resistance of C). If the reactance of L is, say 1,000 ohms, then the voltage across L is .001 × 1,000 = 1 V, or one thousand times the signal voltage! (Similarly across C which has the same reactance at resonance).

The practical resonant circuit (or tuned circuit) is based on just two components—an inductor and a capacitor. Some resistance is always present, however. At low to moderately high frequencies, most of this resistance will come from the wire from which the coil is wound. At very much higher frequencies, the majority of the resistance may come from the frequency energy loss in the capacitor.

TUNED CIRCUITS

The above is the basis for the design of the simplest possible *series tuned circuit.* By making C (or L) adjustable in value it can be tuned (i.e., become *resonant,* with high circuit *magnification*) over a range of frequencies. The usual case is to make C the variable component.

Specifically the problem then becomes one of settling for one component value and then calculating a matching value for the other. Theoretically it does not matter what the 'first' value is as a matching value can always be calculated for the second. In practice capacitance values cannot be reduced indefinitely, and there are disadvantages in making them too large. Practical values for C are therefore chosen consistent with established practice—e.g., 0-250 pf or 0-500 pf for a domestic radio tuned circuit working on the AM bands. The zero value with such components is nominal. It will never be as low as this and could be anything from 10 - 50 pf, actual values being quoted by the manufacturer. To this should be added a further 5 -15 pf to account for capacitance of wiring and other circuit components to arrive at a realistic value for minimum capacitance.

Having selected a particular capacitor one of the following formulas can then be used to calculate the matching inductance required. (These formulas are adjusted for capacity in pF and giving inductance in μH).

$$L = \frac{25,600 \times 10^6}{f^2 \times C}$$

where F = frequency in kHz

$$L = \frac{25,600}{f^2 \times C}$$

where f = frequency in MHz

Example: Find suitable component values for a tuned circuit covering the frequency range 500 kHz to 3,500 kHz.

(i) Make a choice of variable capacitors—say 16 - 250 pF.

(ii) Adjust the lower value for additional capacitance—say making it 16 + 24 = 40 pF as a simple number to work with.

(iii) Use the highest capacity value available to calculate the inductance required to provide the lowest resonant frequency required. (Remember; highest capacity = lowest resonant frequency; lowest capacity value = highest resonant frequency.)

$$L = \frac{2,560 \times 10^6}{(500)^2 \times 250}$$

$$= 40 \ \mu H$$

(iv) Use the lowest capacity value together with the value for L first determined to calculate the highest resonant frequency provided by this combination. For this the formula is rearranged as:

$$f(kHz) = \frac{160,000}{LC}$$

$$= \frac{160,000}{40 \times 40}$$

$$= 4,000 \ kHz$$

This is above the required range, so the component values are satisfactory. If not, there are several alternatives:

(i) Choose a different value variable capacitor and recalculate.

(ii) Recalculate a value for the inductance required to meet the top end requirement. Check whether or not this comes near enough to the bottom end requirement.

(iii) Adjust the working value of the capacitor via a paralleled 'trimmer' capacitor.

(iv) Adjust coil inductance via the core (if practical). Note a tuned circuit whose ultimate performance is adjustable via both a trimmer capacitor and a tunable inductance is called a *double tuned* circuit.

Q FACTOR

The effect of *resistance* in the tuned circuit is shown in simple diagrammatic form in Fig. 26-2, representative of a resonant circuit. The current flowing in a resonant circuit peaks at the resonant frequency and falls off sharply on either side. The lower the *resistance* present, the *higher* the peak (more current flowing) and the sharper it is (the sharper the tuning).

This can be put another way. The shape of the resonant curve is dependent on the respective values of the *reactance* on either the coil or capacitor and the resistance present. The ratio of the two is known as the Q factor, when:

$$Q = \frac{\text{reactance (X)}}{\text{resistance (R)}}$$

Reactance, in ohms, can be calculated from the following formulas:

In the case of an inductance, $\quad XL = 2 \pi fL$
In the case of a capacitor, $\quad\quad XC = 1/2 \pi fL$

At resonant frequency the reactance of a coil and capacitor are the same, as it does not matter which is considered. A simple calculation will prove this, taking the previous values calculated and a resonant frequency of 500 kHz, namely:

$$\text{inductance} = 200 \ \mu H$$
$$\text{capacitance} = 500 \ pF$$
$$\text{Thus } XL = 2 \pi \times 500 \times 10^3 \times 200 \times 10^{-6} \quad = 628 \text{ ohms}$$
$$XC = 1/2 \pi \times 500 \times 10^3 \times 500 \times 10^{12} \quad = 628 \text{ ohms}$$

The *resistance* refers to the dynamic resistance to rf currents in the circuit, not the *dc* resistance. Dynamic resistance is generally known as *impedance*. In the case of a simple air-cored coil, dynamic resistance may rise up to 100 ohms or more, yielding a Q of less than 10. Very much more efficient coils can be produced with Q factors ranging up to 100 (or very much higher in certain cases). These are invariably wound on a non-conducting magnetic core, of ferrite or iron dust bound together with an insulator. The actual value of Q achieved has the same effect as that illustrated in Fig. 26-3. The higher the Q value the sharper, and higher, the peak of the curve.

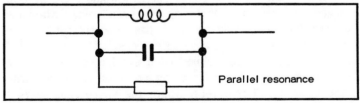

Fig. 26-2. Effect of resistance in a tuned circuit.

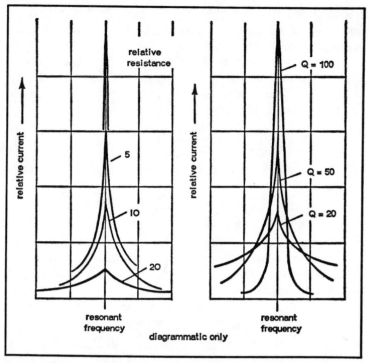

Fig. 26-3. The relationship between resonant frequency and relative values of Q.

With decreasing Q value the tuning becomes *broader* and the peak value is reduced. Sharpness of tuning is always desirable in radio receivers as it gives good *selectivity*, or the ability to separate one station from another when the two are closely spaced on the frequency band.

MODULATED SIGNALS

It is possible to increase the Q attainable from coils to very high levels, using some form of positive feedback to neutralize, or partially neutralize, resistance losses. This would seem an ideal arrangement to get very sharp tuning. However, tuning can be made too sharp for a radio receiver as, it has to accept not a single frequency represented by the *rf* carrier wave, but a whole band of frequencies representing the modulated signal, otherwise some of the *af* content may be 'tuned out' or cut off. This is particularly true in the case of FM receivers, where the signal represents a *bandwidth* rather than a specific *rf* frequency. It is in this respect, both for AM and FM radio that the superhet receiver scores since it changes the *rf* carrier and its sidebands to a single fixed frequency, known as the *intermediate frequency*, and so selectivity can be sharply peaked.

SELECTIVITY AND SENSITIVITY

Selectivity is the ability of a circuit to tune sharply. *Sensitivity* is the ability to amplify the very weak *rf* signals received in the antenna circuit to a practical output level, both as regards signal strength and depth of modulation. It is thus just as important as selectivity for satisfactory receiver performance. Both selectivity and sensitivity increase with increasing Q, but the two are not necessarily compatible.

As a basic example take the simple form of tuned circuit with a tapping point on the coil for connecting directly to the detector stage. The end of the coil connected to the ground of the circuit is referred to as the 'earthy' end. The tapping point on a coil normally comes about one third the length (number of turns) from this end of the coil. If this tapping point is moved up towards the other, or 'hot' end of the coil, this will have the effect of increasing *selectivity* but reducing signal strength or *sensitivity*. Conversely, moving the tapping point towards the 'earthy' end of the coil will reduce selectivity, but increase sensitivity. This, in fact, is one way of adjusting the selectivity and sensitivity of a simple tuned circuit of this type, the aim being to arrive at an optimum tapping point which gives the best possible compromise between selectivity and sensitivity. In practice this does not necessarily mean physically altering the tapping point. With a simple coil it is more practical to add turns at one end and remove turns from the other to 'shift' the tapping point.

Chapter 27

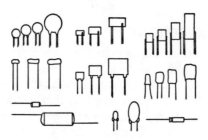

Impedance
Matching Networks

An inductance and a capacitance provides a simple impedance matching network with the two possible configurations as shown in Fig. 27-1. If R_{in} is greater than the load resistance R_L then the inductance L 'follows' the shunt capacitor C and the following design formulas apply:

$$X_L = \sqrt{R_L \times R_{in} - R_L^2}$$

$$X_C = \frac{R_L \times R_{in}}{X_L}$$

Q of circuit $= X_L/R_L$ or R_{in}/X_C

If R_{in} is less than R_L then the inductance 'leads' the shunt capacitor and the following design formulas apply:

$$X_L = \frac{R_L \times R_{in}}{X_C}$$

$$X_C = R\sqrt{\frac{R_{in}}{R_L - R_{in}}}$$

Q of circuit $= X_L/R_{in}$ or R_L/X_C

The respective inductive reactance and capacitive reactance required can thus be calculated, and from these corresponding values for L and C remembering:

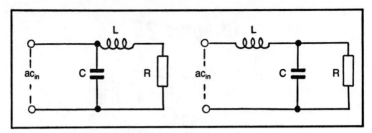

Fig. 27-1. L-networks (impedance matching circuits).

$$X_c = \frac{1}{2 \pi fC}$$

$$\text{or } C = \frac{1}{2 \pi fX_c}$$

where f is in Hz
and C is in farads

$$X_L = 2 \pi fL$$

where f is in Hz
and L is in henrys

The above are known as L-networks.

Chapter 28

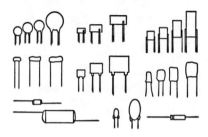

Multivibrators

The basic multivibrator is a cross-linked circuit between two transistors both of which are working in an unstable state so that the circuit oscillates between the two states. This is initiated by slight imbalance in the components, or random variations in circuit, so the circuit is self-starting. The basic circuit is shown in Fig. 28-1. The time for which transistor T1 is cut off and T2 is saturated is given approximately by:

$$t_1 = C1R1 \log 2$$

Similarly, for the other half cycle where T1 is saturated and T2 is cut off:

$$t_2 = C2R2 \log 2$$

The following simple design rules also apply for determining suitable component values.

Make R1 more than $\frac{1}{2}\beta_1$ R2
Make R4 more than $\frac{1}{2}\beta_2$ R3

Ensure that the base-emitter voltage V_{be} of the transistor is much lower than the supply voltage. Further to simplify the circuit design:

Make R1 = R4
R2 = R3
C1 = C2

T1 and T2, of course, should also be similar transistors. Time for each oscillation can then be taken as:

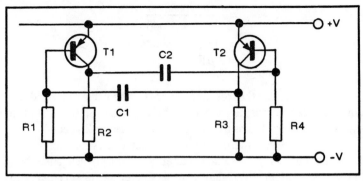

Fig. 28-1. Basic multivibrator circuit.

$$t = 2 \times C1\ R1\ \log 2 \text{ seconds}$$
$$= 0.7\ C1\ R1 \text{ seconds (approx.)}$$

Example: Suppose the transistors chosen are two similar low power germanium pnp types with a β of 50 and an I_c max of 10 mA.

(1) 'Guesstimate' a suitable supply voltage—say 9 volts.

(ii) Calculate a value for R1 based on limiting I_c to less than 10 mA.

$$I_c = V/R1$$

$$\text{or } R1 = \frac{V}{I_{c\ max}}$$

$$= \frac{9}{.010}$$

$$= 900 \text{ ohms}$$

Make R1 = 1 k ohm
also R4 = R1 = 1 k ohm

(iii) R1 is to be more than ½ × 50 × R2:

i.e., more than 25 × R2
(say)   30 × R2

Then R2 = 30 × 1
= 30 k ohms

Nearest preferred value is 27 k ohms (R2 = R3 = 27 k ohms).

(iv) Calculate a value for C1 for required frequency of oscillation, say 5 kHz. This is equal to a cycle time of:

$$1/5000 = .0002 \text{ seconds}$$

$$C1 = \frac{t}{0.7\ R1}$$

$$= \frac{.0002}{0.7 \times 1,000}$$
$$= .0000002 \text{ farads}$$
$$= 2 \ \mu\text{F}$$

So, make C1 = C2 = .1 μF

Note: To lower the frequency, increase the value of C1 and C2. For example, increasing the value of C1 and C2 to 100 μF in the same circuit would give a cycle time approaching 2 seconds.

Here, however, there is a possible danger as capacitor values of a high order can share sufficient energy to damage the transistors. In the case of *planar* transistors, for example, this is very likely to occur when using values for C1 = C2 in excess of 1 μF.

The generally accepted method of adjusting the timing of a multivibration circuit by altering the capacitor values must therefore be approached with ease to ensure that the reverse V_{eb} rating of the transistors cannot be exceeded. Two treatments are possible (apart from recalculating suitable resistor values to suit):

(i)   Reduce the supply voltage to a 'safe' figure.

(ii) Insert diodes in each emitter lead to ensure complete discharge of the capacitors on each half cycle.

ONE-SHOT MULTIVIBRATOR

The one-shot multivibrator or 'flip-flop' (also known as a monostable vibrator) is basically an ordinary multivibrator in which

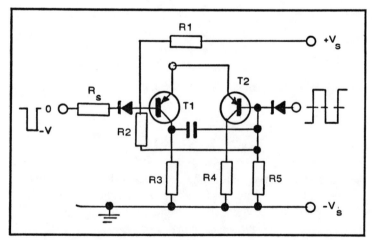

Fig. 28-2. One-shot multivibrator. Component values can be worked out to match transistors chosen (see text).

one coupling is direct. As a result it is triggered by a pulse from one state to the other where it remains quiescent, until triggered back again by the next input pulse to a quiescent state again. A basic circuit is shown in Fig. 28-2.

Design Requirements

The basic design requirements are:

(i) In the quiescent state T2 is saturated and T1 is cut off.

(ii) In the triggered state T1 is saturated by the current flowing through R2 and R4 even if the input is removed (i.e., the triggering pulse has ceased).

Step-By-Step Design

(i) R3 or R4 represents the load and for similar transistors can be made equal, i.e., R3 = R4.

(ii) Calculate R5 as equal to $\frac{1}{2}\beta$ R4.

(iii) Calculate R1 to allow for saturation in the triggered condition, e.g., take current through R1 as 50 times 1_{cbo} for the transistor. (Take number of I_{cbo} as 0.5 mA for a germanium transistor and 1 μA for a silicon transistor).

(iv) Determine current in T1 when bottomed and from this base current required to ensure bottoming.

(v) (R2 + R4) must supply base current for bottoming plus V_s + R1.

(vi) Determine R2 from the relationship:

$$R2 + R4 \leq V_5/\text{current from (V)}$$

(vii) Calculate the value of C from t = 0.7 CR5

Design example: Supply is + 5 V and –10 V. Load is 2.2 k ohm. Flip-flop time is to be 1 millisecond. Silicon transistors are chosen for the circuit having a β of 30 at 25 mA.

(i) R3 – R4 load = 2.2 k ohm
(ii) R5   = $\frac{1}{2}\beta$ × R4
                = $\frac{1}{2}$ × 30× 2.2
                = 33 k ohm

This is a preferred value and needs no adjustment.

(iii) Current through R1 is to be 50 × .000001 = .00005 mA.

$$\text{Here, R1} = \frac{5}{.00005}$$
$$= 100 \text{ k ohms}$$

(iv) Current in T1 when bottomed is $10/2.2 = 4.5$ mA, thus base current for bottoming should be greater than $4.5 \times (2/\beta) = 4.5 \times 1/15 = 0.3$ mA.

(v) R2 + R4 in series must now supply:

$$0.3 \text{ mA} + V_s +/R1$$
$$= 0.3 + 5/100$$
$$= 0.3 + .05$$
$$= .35 \text{ mA}$$

(vi) R2 + R4 $\leq 10/.35$
$$\leq 28.5 \text{ k ohm}$$

But R4 has already been established as 2.2 k ohm.

$$\text{thus, R2} = 28.5 - 2.2$$
$$= 26.3 \text{ k ohm}$$

Adjust to nearest preferred value:

$$\text{R2} = 27 \text{ k ohm}$$

(vii) Rewriting the formula for t as a solution for C

$$C = \frac{t}{0.7 \times R5}$$
$$= \frac{.001}{0.7 \times 33,000}$$
$$= .043 \ \mu F$$

adjust to nearest preferred value

$C = .039 \ \mu F$ or $.047 \ \mu F$

Note: adjusting to the nearest preferred value will, of course, alter the flip-flop time. If this is a critically important feature, recalculate a value for R5 using one of the preferred values for C, say $.039 \ \mu F$.

Then for t to remain at .001 seconds:

$$R5 = \frac{.001}{40.7 \times .039}$$

$$= 36.6 \text{ k ohm (nearest preferred value 39 k ohm)}$$

Try with $C = .047 \ \mu F$

$$R5 = \frac{.001}{0.7 \times .047}$$

$$= 30.4 \text{ k ohm (nearest preferred value 33 k ohm)}$$

Neither is a much better solution, so stay with the original calculation.

IC MULTIVIBRATORS

Design of multivibrators from scratch using discrete components is, of course, the long way round. You can get the same results, with no hassle, simply by using an IC package. Here, of course, external component values to be used are tied to the specific IC employed.

The simplest form of IC multivibrator merely uses an op amp in a basic oscillator circuit such as that shown in Fig. 28-3. Oscillation frequency will depend on the IC parameter and the values of the external resistors. With the components shown, the output frequency will be 1 kHz and in the form of a square wave. With the addition of a diode to this circuit the pulse width can be adjusted by using different values for R2. The value of resistor R3 governs the actual pulse duration (Fig. 28-4).

An alternative form of multivibrator is to use two op amps connected as cross-coupled inverting amplifiers, as shown in Fig. 28-5. Here the frequency is established by the time constants of the RC combination R1C1 and R2C2. R1 and R2 should be the same value, and can be anything from 1 k ohm 10 k ohms. C1 and C2 should also be similar values, and anything from 0.01 to 10 μF can be used. The basic rules governing adjustment and oscillation frequency are that for any particular value of R1 and R2, *increasing* the value of C1 and C2 will *decrease* the oscillation frequency, and vice-versa. Similarly, for any particular value of C1 and C2 *decreasing* the value of R1 and R2 will *increase* the frequency, and vice versa.

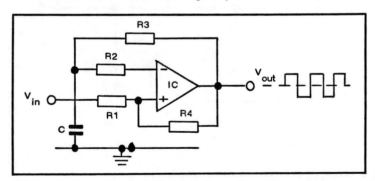

Fig. 28-3. IC multivibrator using CK3401 op amp. Component values: R1 – 10 M ohm; R2 3 M ohm; R3 – 30 k ohm; R4 – 10 M ohm; C – .01 μF.

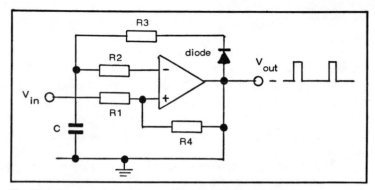

Fig. 28-4. The same circuit improved by the addition of a diode.

With the component values shown, i.e., R1 = R2 = 8.2 k ohms and C1 = C2 = 0.2 μF, the oscillation frequency will be 1 kHz. Decreasing the value of R1 and R2 to 1 k ohm should result in an oscillation frequency of 10 kHz.

A rather more versatile multivibrator circuit is shown in Fig. 28-6, which has independent controls of 'on' and 'off' periods. The frequency range if adjustable by choice of capacitor C1 which governs the duration of the square wave pulse generated, viz:

Value of C1	pulse period	frequency
1F	4 min to 1 sec	250 - 1 Hz
0.1 F	0.4 min to 100 min	2,500 - 600 Hz
0.01 F	4 min to 10 min	1,500 - 6,000 Hz
0.00 F	4 sec to 1 min	15,000 kHz - 60 kHz

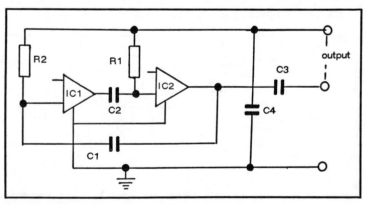

Fig. 28-5. Multivibrator circuit using two μL914 op amps. R1 and R2 can be anything from 1 k ohm to 10 k ohm. C1, C2 and C3 — .01 μF; C4 — 100 μF. Supply; 3.6–6 volts.

234

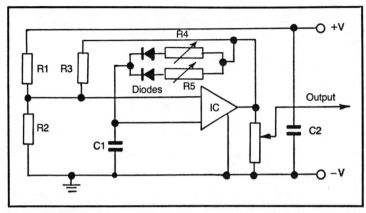

Fig. 28-6. Multivibrator circuit with adjustable 'on' and 'off' periods via the two 1 M ohm potentiometers. Other component values: R1, R2 and R3 – 100 k ohms; R4 and R5 – 1 M ohm; R6 and R7 – 2 k ohms. Capacitors: C1 – selected as required; C2 – .01 µF. IC-CA3130. Supply: 15 volts.

Adjustment of 'on' and 'off' times of oscillation within these ranges is governed by the potentiometers R4 and R5.

Another multivibrator circuit is shown in Fig. 28-7 which is particularly notable for its stable performance. The frequency of oscillation is maintained to within plus or minus 2 percent on any supply voltage from 6 to 15 volts and is independent of the actual voltage. It uses a CA3094 op-amp IC with external resistors and one capacitor. The circuit also includes a lamp which flashes on and off at a rate of one flash per second with the component values given.

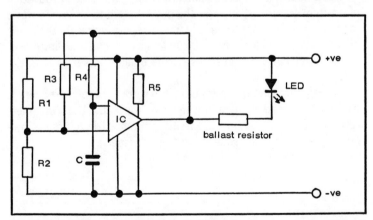

Fig. 28-7. Multivibrator circuit working a flashing light. Component values: R1 – 3 M ohms; R2 – 12 M ohms; R3 – 16 M ohms; R4 – 4.3 M ohms; R5 – 1.2 M ohms; C – .45 µF; IC – CA 3094. Supply: 6 to 12 volts.

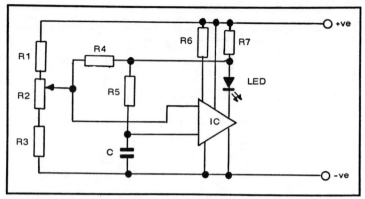

Fig. 28-8. Adjustable multivibrator/flashing light circuit. Component values for flashing rate of approx. 1 sec: R1 27k ohm; R2 50 ohm pot; R3 −27 k ohms; R4 and R5 −100 k ohms; R6 − 300 k ohms; R7 − ballast resistor; C − 500 pF; IC − CA309A. Supply = 16½-25 volts.

Flashing rate can be adjusted by altering the values of R1 and R2 and/or C. To adjust values to give any required flashing rate (frequency), the following formula applies:

$$\text{frequency} = \frac{1}{2RC1_n(2R1/R2 + 1)}$$

$$\text{where } R1 = \frac{RA \cdot RB}{RA + RB}$$

In a variation on this circuit shown in Fig. 28-8, the introduction of a potentiometer R2 enables the pulse length to be varied while maintaining a constant frequency (pulse repetition rate). Again this circuit can be used to flash a filament lamp, or a light emitting diode. In the latter case, a ballast resistor is needed in series with the LED.

Designing a multivibrator circuit to work at an audio frequency, while retaining adjustment of frequency, forms the basis of a metronome. The only additional circuitry required is a simple low-power audio amplifier connecting to a loudspeaker.

Chapter 29

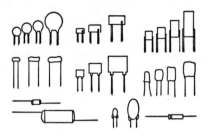

Amplifier Design

An amplifier is a device (consisting of a suitable circuit) which accepts small signal inputs and generates magnified (amplified) output signals. The basic parameters involved are thus:

(i) The input signal to be accepted—i.e. its volume and value(s).

(ii) The actual circuit producing the amplification.

(iii) The output signal value(s) and also its characteristics (e.g. distortion and stability).

As far as the design of an amplifier circuit is concerned, it is best to start the other way round. That is, first establish specific output requirements, design the output stage accordingly. This will also establish the signal levels required to drive it, which sets the requirements for any previous stage, and so on back to the input. In other words design is worked back from output through to input.

SIMPLE POWER AMPLIFIERS

A basic requirement in any transistor power amplifier is a stable operating point, otherwise there will be very real danger of thermal runaway with the transistor being destroyed. In practice, this simply means a suitable bias circuit to 'hold' the dc operating point of the transistor steady. While this subject has already been raised in the chapter on *basic transistor circuit design* (Chapter 10), some repetition is included here as an essential feature of power amplifier design.

The most common type of biassing circuit used is that shown in

Fig. 29-1, using two base resistors R1 and R2 to determine the value of the input voltage V_{BE} in conjunction with the negative feedback from the emitter resistor R_e. Any increase in emitter current causes a voltage drop across R_e, which reduces the base-emitter voltage (V_{BE}). Such compensation for change is thus automatic.

The actual *stability* of such an arrangement is defined by the stability factor K. The greater the value of K the better the stability, so it is very useful to be able to calculate K directly. The formula is:

$$K = 1 + \frac{h_{fe} \times R_e}{R_R + R_B}$$

$$\text{where, } R_B = \frac{R1 \times R2}{R1 + R2}$$

What this means is that the larger the value of the emitter resistor R_e compared with R_B the higher the value of K, and hence the greater the stability of the circuit.

There are limits to how *large* R_e can be made, dictated by how much of the supply voltage can be dropped across it and consequently how much voltage is available as signal swing in the collector resistor R_e.

Equally there are limits as to how *small* R1 and R2 can be made, depending on how much current may be drawn from the power supply by them. Also, if R1 and R2 values are too low they will act as a shunt across the ac input.

At this point we need a rule-of-thumb guideline as a starting point for working out realistic resistor values. This is:

take R_B as equal to $10 \times R_e$

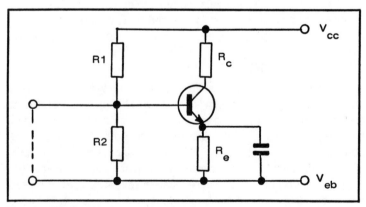

Fig. 29-1. Basic transistor amplifier circuit.

With one other formula to work with we can then set about the design of the complete circuit. These formulas are:

$$V_B = \frac{V_{CC} \times R2}{(R1 + R2)} = V_E$$

Design—Step-By-Step

(i) Select a suitable transistor capable of providing the required power output; and for which the characteristics are known.

(ii) Plot the load line on the transistor output characteristics graph (see Chapter 10).

(iii) Select a suitable operating point.

(iv) Check (a) swing, (b) output power and distortion.

(v) Establish the minimum voltage to which the collector falls (V_C min). From this calculate a suitable value for R_e to make V_E a little less than V_C min.

(vi) Taking R_B as $10 \times R_e$, calculate values for R1 and R2, using known voltage and current values.

(vii) Readjust the theoretical resistor values to the nearest practical values.

(viii) Check current drain on input side and compare with output.

(ix) Calculate the circuit stability factor. Here it may be something of a 'guesstimate' as to what value is needed. The more stability you are aiming at, the higher this value should be.

(x) Confirm, or recalculate, resistor values.

(xi) Determine a suitable value for the decoupling capacitor C_E consistent with the signal frequency involved on the basis that the *reactance* of this capacitor must be small compared with the value of R_e.

Design Example 1—Straightforward Power Amplifier

Requirements: 2 watt output into a load of 10 ohms with low distortion. Input is 24 volts ac with a 30 mA peak-to-peak swing.

(i) Select a transistor with a suitable output power rating—preferably at least five times the output power as low distortion is called for. A suitable choice would have output characteristics like that shown in Fig. 29-2, with a 10 or 15 watt rating.

(ii) Plot the load line:

$$\text{at } V_{CE} = 0 , \ I_C = V_{CC}/R_L$$
$$= 24/10$$

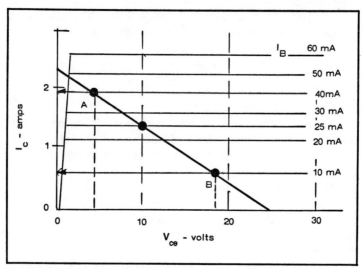

Fig. 29-2. Characteristics of suitable transistor with working point(s) established.

$$= 2.4 \text{ amps}$$
$$\text{at } I_C = 0 \text{ , } V_{CC} = 24 \text{ volts.}$$

(iii) Select a suitable working point—say at $V_{CE} = 10$ and $I_C = 14$, corresponding to a base current I_B of 25 mA. (All these values read off transistor characteristic curves.

(iv) (a) Current 'swing' of 30 mA peak to peak means a 15 mA swing in base current about 25 mA—i.e., from 40 mA down to 10 mA (point B on the load line). This results in a swing in collector current between 0.65 and 1.95 amps, or 1.3 amps; with output voltage varying between 4 and 17, and 13 volts 'swing' (again these values are read off the graph).

(b) *Power developed* in the output load is equal to current swing multiplied by voltage variation, divided by 8:

$$\frac{1.3 \times 13}{8} = 2.115 \text{ watts}$$

This is confirmed as satisfactory.

(c) Distortion can be estimated on the basis of the ratio of the lengths A0 and 0B on the load line. If A0 = 0B there is no distortion.

(v) The minimum value of the collector voltage V_C has already been established at (iv)(a) and V_C min = 4V. The emitter voltage needs to be a little less than V_C min—say 3.5 volts.

The standing collector current at the working point is 1.4 amps.

$$\text{Hence } R_e = \frac{\text{emitter voltage}}{\text{steady current}}$$

$$= \frac{3.6}{1.4}$$

$$= 2.5 \text{ ohms}$$

Note: this is not a *real* (practical) resistor value. Nearest available values are 2.2 or 2.7. We can continue with a theoretical value (and recalculate as necessary later), or adjust now. The latter treatment is best, but we need to decide which way to go.

Taking R_e as 2.7 ohms gives an emitter voltage = 1.4 × 2.7 = 3.78. Taking R_R as 2.2 ohms gives an emitter voltage = 1.4 × 2.2 = 3.08. The latter is probably the safer choice (3.78 volts is quite close to V_{Cmin} of 4 volts).

(vi) Having established a value of 2.2 ohms for R_e, we can now say:

$$R_B \text{ is to equal } 10 \times 2.2 = 22 \text{ ohms}$$

$$\text{or } 22 = \frac{R1R2}{R1 + R2}$$

We now have to work with the two formulas given previously, viz:

$$V_B = \frac{V_{cc} \times R2}{(R1 + R2)}$$

Substituting known values:

$$3.08 = \frac{24 \times R2}{(R1 + R2)}$$

i.e. $R1 + R2 = 8R2$ (approx.) or $R1 = 7R2$.

Substituting in the first calculation

$$22 = \frac{7R2 \times R2}{8R2}$$

$$= \frac{7R2}{8}$$

$$\text{or } R2 = 25 \text{ ohms}$$

$$\text{when } R1 = 7R2 = 175 \text{ ohms}$$

(vii) Again these need adjusting to practical values. To keep the total resistance on the input side down, take the nearest higher values:

$$R2 = 27 \text{ ohms}$$
$$R1 = 180 \text{ ohms}$$

(viii) On the input side of the circuit R1 and R2 are in series, giving a total resistance of $27 + 180 = 207$ ohms. Input voltage is 24 V, so current drain on input side is:

$$\frac{24}{207} = 116 \text{ mA}$$

This is quite small compared with the standing current of the transistor (1.4 amps) and so is fully acceptable.

(ix) Calculate the stability factor from:

$$K = 1 = \frac{h_{fe} \times R_E}{R_R + R_B}$$

First we have to recalculate R_B as the final values of R1 and R2 have been modified.

$$R_B = \frac{R1 \times R2}{R1 + R2}$$
$$= \frac{180 \times 27}{180 + 27}$$
$$= 23.5 \text{ ohms}$$

We also need to know the gain (h_{fe}) of the transistor being used. Say this is 50.

$$\text{Then, } K = 1 + \frac{50 \times 2.2}{2.2 + 23.5}$$
$$= 1 + 4.28$$
$$= 5.28$$

This is a fairly low value representing moderate stability. If a higher stability factor is required, resistor values need to be adjusted.

(x) Recalculation of resistor values. The simplest way of doing this is to calculate the value of R_B for a *given* stability factor. Calculation for R_e is *not* affected and the original value determined is used. Suppose a stability factor of 10 is called for, then, for the same transistor:

$$10 = 1 + \frac{50 \times 2.2}{2.2 + R_B}$$
$$\text{or } R_B = \frac{110 - 19.8}{9}$$
$$= 10 \text{ ohms}$$

R1 and R2 are now recalculated on the same basis as before:

$$R1 = 7R2 \text{ (as before)}$$

$$R_B = \frac{R1 \times R2}{R1 + R2}$$

$$\text{or } 10 = \frac{7R^2}{8}$$

$$\text{where, } R2 = 11.42 \text{ ohms}$$
$$\text{and, } R1 = 7R2 = 80 \text{ ohms}$$

Adjusting to practical values:

$$R1 = 82 \text{ ohms}$$
$$R2 = 12 \text{ ohms}$$

This will, of course, slightly modify the stability factor.

Adjusted value of R_B is $\dfrac{82 \times 12}{82 + 12} = 10.47$ ohms

Adjusted value of K is $1 + \dfrac{50 \times 2.2}{2.2 \times 10.47} = 9.68$

The difference is not worth worrying about. If it was a bigger difference, then reselect the working values of R1 and R2.

(xi) The decoupling capacitor value is not likely to be critical. The main requirement is that its *reactance* should be small compared with the value of the emitter resistor R_e - say not more than one-fifth of its value.

$$\text{Reactance (ohms)} = \frac{1}{fC}$$

when f is the frequency in Hz
and C is the capacity in farads

Since reactance is usually proportional to frequency, the important frequency to consider is the minimum frequency. Suppose this is 2 kHz in the case of the worked out example whence the value of R_e has been determined as 2.2 ohms. We want the reactance of the decoupling capacity to be 2.2/5 = 0.44 ohms or less—say 0.4 ohms as a maximum. This will then give a *minimum* capacity value of:

$$C = \frac{1}{2,000 \times 0.4}$$
$$= .00125 \text{ farads}$$
$$\text{or } 1,250 \ \mu\text{f.}$$

AUDIO AMPLIFIER DESIGN

A class A audio (output) amplifier employs an identical circuit to that just described, except for choice of transistor and the collector load (R_C) formed by the primary of an output transformer. This means that all component values have to be recalculated around the specific transistor selected (i.e., an audio amplifier type). Before this, it is necessary to examine the output and power requirements.

Effectively the operating point of the transistor can range over the whole length of the load line contained within the characteristic output curve(s)—i.e., above the 'knee' voltage (Fig. 29-3). The design operating point is then half way along this band line, giving a standing current of I_c with a supply voltage of V_{cc}. It is then possible for the collector voltage to swing between the knee voltage and $2 \times V_{cc}$. Similarly the collector current can swing from zero to twice the standing current ($2 \times I_c$).

The *load* represented by this load line is equal to (V_{cc}—knee voltage) divided by the standing circuit I_o, yielding the very useful formula:

$$\text{load resistance (required)} = \frac{(V_{cc} - V_{knee})^2}{2P_o}$$

where P_o is power output

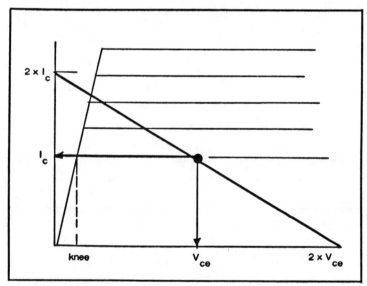

Fig. 29-3. Design operating point for a Class A amplifier.

Design—Step-by-Step

(i) Start by evaluating the load, allowing for power losses (uprate maximum output specified by 10 percent. Adjust design knee voltage value upwards to minimize distortion.

(ii) Establish the power requirements and power rating for the transistor (remembering that a Class A amplifier has an efficiency of only 50 percent).

(iii) Plot the load line for the design load. Establish standing collector current and current and voltage swing.

(iv) Establish end parts A and B on the load line. Recheck the power output. Find a suitable emitter voltage.

(v) Calculate the value of R_e. Adjust to a preferred value.

(vi) Determine R_B as $10 \times R_e$ and from this work out values for R1 and R2.

(vii) Determine a suitable value for the decoupling capacitor C_E.

(viii) Calculate transformer ratio required.

Design Example 2—Class A Audio Amplifier

Requirements: Maximum output of 100 mW into a 4 ohm speaker for a small radio receiver with a supply voltage of 9 V. (Output characteristics of a likely transistor to use are given in Fig. 29-2.)

(i) V_{cc} is given as 9 V. Knee voltage is 0.25 (typical of most af silicon transistors). Uprate maximum output required by 10 percent = 110 mW. Calculate the load required:

$$R_L = \frac{(V_{cc} - V_{knee})^2}{2 \times 110 \text{ mW}}$$

$$= \frac{(9 - 0.25)^2}{2 \times .11}$$

$$= 348 \text{ ohms}$$

It is worth working to a higher knee voltage to minimize distortion? Let's say it is, and raise the knee voltage to 1.0 V.

$$\text{Now, } R_L = \frac{(9 - 1)^2}{2 \times .11}$$

$$= 290 \text{ ohms}$$

Settle for one or the other—or something in between—say R_L = 300 ohms

(ii) We are dealing with a maximum output power of 100 mW.

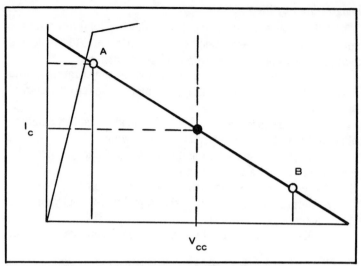

Fig. 29-4. Design operating point from calculated example.

Multiply by 2.5 for a *minimum* power rating for a suitable transistor—250 mW. If the transistor previously selected is below this, reselect one which has a power rating of *at least* 250 mW.

(iii) Plot the load line for design load ($R_L = 300$ ohms) on the transistor characteristics.

$$\text{At } V_{CE} = 0, \; I_C = V_{CC}/R_L$$
$$= 9/300$$
$$= 30 \text{ mA}$$
$$\text{At } I_C = 0, \; V_{CR} = 2 \times V_{CC} = 18 \text{ volts}$$

Superimposed on the transistor characteristic (Fig. 29-4), this will establish the design operating point.

(iv) The load resistance formula can be rearranged as a solution for power output:

$$P_o \text{ (max)} \quad = \frac{(V_{CC} - V_{min})^2}{2 \times R_L}$$

Establishing V_{min} a little higher than the knee voltage, say 1.4 V:

$$P_o \text{ (max)} \quad = \frac{(9 - 1.4)^2}{2 \times 300}$$
$$= 96 \text{ mW}$$

This is near enough to the original figure, so we can use this to establish the emitter voltage as a little less than 1.4 V—say 1 V.

(v) The steady current at the working point has been established by superimposing the load line on the transistor output curves. Say this is 20 mA:

$$R_E = \frac{\text{emitter voltage}}{\text{steady current}}$$

$$= \frac{1}{.020}$$

$$= 50 \text{ ohms}$$

The nearest preferred value is 47 ohms.

(vi) $R_B = 10 \times R_E$
$= 10 \times 47$
$= 470 \text{ ohms}$

$$\text{or } 470 = \frac{R1 \ R2}{R1 + R2}$$

$$\text{also, } V_B = \frac{V_{cc} \times R2}{R1 + R2}$$

$$\text{i.e., } 1 = \frac{9 \times R2}{R1 + R2}$$

where, $R1 = 8R2$

Substituting in the first formula:

$$470 = \frac{8R2^2}{9R^2}$$

$$\text{or } R2 = 528 \text{ ohms}$$

where, $R1 = 4,230 \text{ ohms}$

Readjusting to nearest preferred values:

$R1 = 3.9 \text{ k ohms.}$
$R2 = 560 \text{ ohms.}$

(vii) Total resistance on the input side is $560 + 3,900 = 4.46$ k ohms. Input voltage is 9 V, so current drain on input side is:

$$\frac{9}{4,460} = 2 \text{ milliamps}$$

This is small compared with the standing current of the transistor (30 mA) and so quite acceptable.

(viii) In step (i) we established the primary resistance of the transformer as 300 ohms. The speaker represents an output load of 4 ohms. Hence for matching:

$$\text{Transformer turns ratio required} = \sqrt{\frac{\text{primary resistance}}{\text{output load resistance}}}$$

$$= \sqrt{\frac{300}{4}}$$

$$= 8.66:1$$

(ix) Since this is an audio circuit the minimum frequency will be quite low—say 40 Hz. The value of R_E is 50 ohms, so the capacitor should have a reactance of no more than $50/5 = 10$ ohms. This will give a minimum capacitor value of:

$$C = \frac{1}{40 \times 10}$$

$$= 250 \ \mu F$$

CLASS B AMPLIFIER DESIGN

The immediate advantage offered by a push-pull class B amplifier is that the output power obtained is considerably greater than double the power of a single transistor. Also the average current drain is very much lower than with Class A operation because the transistors are biased so that their working point is near cut-off and quiescent current is virtually zero. (Figure 29-5 shows a 'working' diagram of Class B operating characteristics).

Apart from much lower current consumption, making it more suitable for battery operated circuits, the efficiency of a Class B output can approach 80 percent. It does have its inherent limitations, however, and in particular a proneness to *crossover distortion*. This is distortion produced at the change-over point when working is transferred from one transistor to another (changing from 'push' to 'pull' and 'pull' to 'push').

Crossover distortion will be most marked if both transistors are biased exactly to cut-off. It can be overcome, or at least the residual distortion can be substantially reduced, by selecting the bias so that, one transistor does not cut-off until the other has stopped conducting i.e., there is a slight overlap at the changeover. Unfortunately the amount of overlap will tend to change both with the operating temperature of the transistors and any change in the supply voltage. Equally, differences in the spread of characteristics of different transistors of the same type can make design for optimum bias difficult or even impossible without further 'cut and try' adjustment of values. It is possible to incorporate compensating components in the circuit design to minimize the undesirable effects of temperature and characteristic spread. If necessary the supply voltage can also be stabilized (e.g., by means of a zener diode).

A standard circuit for a Class B amplifier is shown in Fig. 29-6. Crossover distortion is eliminated by supplying the drive from a high resistance source and applying a small forward bias to each transistor via R1 and R2. A typical value of bias used is 100-200 mV. Preferably R2 should be shunted with a thermistor to stabilize the effective resistor value under temperature changes. The emitter resistor R_R also acts as a stabilizing device.

Design—Step-by-Step

(i) Select a suitable *matched pair* of transistors. These should be capable of dissipating one fifth of the peak power output.

(ii) Calculate the load requirements from:

$$R_c = \frac{V_{cc}^2}{2 P_{o(max)}}$$

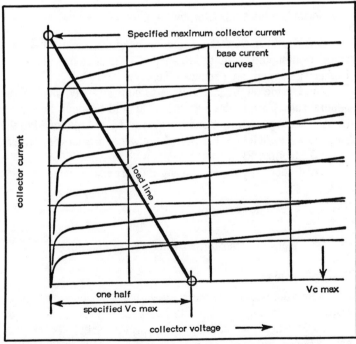

Fig. 29-5. Load line superimposed on transistor characteristics for Class B amplifier.

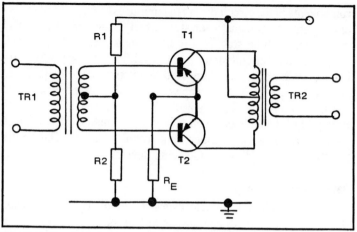

Fig. 29-6. Standard circuit for a Class B amplifier.

$$R_c = \frac{V_{cc}^2}{3\, P_{o(max)}}$$

This formula allows for 'overpowering' by 50 percent.

(iii) Establish a suitable value for the emitter resistor R_E. The value used should be quite low—say 4.7 ohms as a general rule. Higher values will only reduce efficiency.

(iv) Decide on the bias voltage to be used—say 150 mW as a general rule. Current drain should be limited to about the level of the quiescent current in transistors—say 1.5 mA. Proceed to find suitable values of R1 and R2. Note these values are not very critical.

(v) Stabilize R2 with a thermistor?

(vi) Check output power using the formula:

$$\text{Useful power output} = P_{o(max)} \times \frac{R_c}{R_c + R_E}$$

(vii) Work out transformer turns ratio required.

Design Example

Required: Class B output stage with a peak power output of 1 watt. Supply voltage available 12 V. Speaker resistor is 8 ohms.

(i) Matched pair of transistors (provisionally) chosen must have a power rating of at least 1/5 watt, i.e., at least 200 mW.

(ii) Load presented to each collector is:

$$R_C = \frac{V_{cc}^2}{3P_{o\ max}}$$

$$= \frac{12^2}{3 \times 1}$$

$$= 48 \text{ ohms}$$

(iii) Value of emitter resistor R_E is to be 4.7 ohms.

(iv) Resistors R1 and R2 are in series, limiting current drain to the desirable value of 1.5 mA with a given supply voltage of 12 V.

$$\text{Hence R1} + \text{R2} = \frac{12}{1.5 \text{ mA}}$$

$$= 8,000 \text{ ohms}$$

Bias voltage is to be 150 mV. Resistors R1 and R2 act as a voltage divider with the required bias voltage dubbed from 12 V to 150 mV across R1. This means that the ratio of R2/R1 required is 12/.150 = 80 or:

$$\text{R1} = 79 \text{ R2}$$

Substituting for R1 in the original equation:

$$80 \text{ R2} = 8,000 \text{ ohms}$$
$$\text{or, R2} = 100 \text{ ohms}$$
$$\text{when, R1} = 79 \times 100$$
$$= 7.9 \text{ k ohms}$$

Suitable final preferred values are then:

$$\text{R1} = 8.2 \text{ k ohms}$$
$$\text{R2} = 100 \text{ ohms or } 120 \text{ ohms}$$

(v) If R2 is to be stabilized against change of resistance with temperature or negative temperature coefficient thermistor must be connected in parallel with it. Since the two are resistances in parallel, the value of R2 and the thermistor must be twice the original figure calculated in step (iv), i.e., 200 ohms or 240 ohms for both R2 and its parallel thermistor.

(vi) V_{cc} is 9 volts. R_C has been determined in step (ii) as 48 ohms. R_E has been fixed at 4.7 ohms. Design output used in determining R_C was 50 percent over requirement, i.e., 1.5 watts.

$$\text{useful power output} = 1.5 \times \frac{48}{48 + 4.7}$$

$$= 1.37 \text{ watts}$$

(vii) Each transistor sees *half* the full load for *half* the time. The effective collector load is thus 4 times R_c.

$$\text{collector load} = 4 \times 48$$
$$= 192 \text{ ohms}$$

Speaker resistance is 8 ohms.

Hence transformer turns ratio to match is $\sqrt{\dfrac{192}{8}} = 4.9$

Chapter 30

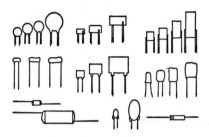

Constant Current Circuits

Constant current circuits are commonly a necessity in time bases, signal processing equipment, integrators, etc. An op-amp can provide a simple solution, requiring only a minimum of external components (normally all resistors). However they have distinct limitations. For example at operating frequencies greater than a few kilohertz, common op-amps suffer from bandwidth and slow rate limitations, general spurious oscillations, or even become inoperative. In such cases it is necessary either to go to a faster (and more expensive) type of op-amp and also possibly need to include fairly complex frequency stabilization networks. In such cases, construction of constant current circuits from discrete components can be simpler, and better.

The simplest of all constant current circuits is a resistor in series with the voltage source, as shown in Fig. 30-1. This is a 'nominally' constant current device only. It depends on a constant voltage source and restriction of current drift by suitable choice of R1 compared with the load resistance R_L. Specifically, the current flowing through the load is given by:

$$\text{load current} = \frac{V_{in} - V_{out}}{R1}$$

To be effective V_{in} must be very much larger than V_{out}, i.e., a high source voltage is available and only a low output voltage is required. Since V_L is determined by the load resistance R_L this equally means that R1 must be very much greater than R_L. V_{out} is

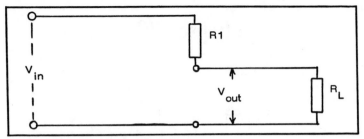

Fig. 30-1. Simple constant current circuit.

'adjusted' to the working value required by choosing R1 to drop the output voltage to that required to give the required current through the load R_L, i.e., to make:

$$V_L = R_L \times \text{current required}$$

The departure from constant current characteristics in the output circuit is then given by:

$$\text{departure} = \frac{100 \times V_L}{V1} \text{ percent}$$

For example, to hold the output circuit within 1 percent departure from the specified current value required

$$1 = \frac{100 \times V_L}{V1}$$

$$\text{or } \frac{V1}{VL} = 100$$

Thus for a design output of 2 volts, for example, R1 would need to be 100 times R_L, with a voltage source of 200 V.

Apart from the disadvantage of requiring a high source voltage, the other virtue of having V1 very much larger than V_L is that any variation in V_L will have very little effect on modifying the output circuit current, since this is directly proportional to (V1–V_L). If V1 is 200 V and V_L is 2 V, for example, and the load resistance is 100 ohms, the stable load current is 2/100 = 20 mA. If the load resistance sources have to swing V_L to, say 1.95 volts, the departure in current value is only (200 – 2) compared with (200 – 1.95) or 198/198.05 or a matter of 0.025 percent.

CONSTANT CURRENT FROM LOW VOLTAGE SUPPLY

The snag with the previous circuit is, of course, the need for a high voltage supply to achieve a large V_{in}/V_{load} ratio for stability.

Where only a low source voltage is available, a transistor circuit as in Fig. 30-2 can provide the answer.

Here R2 and R3 act as a potential divider, holding the transistor base at a voltage equal to $V_{in}/R3(R2+R3)$. The emitter of the transistor will be a few hundred millivolts more positive. The current in R1 is then:

$$\text{emitter current} = \frac{V_{in} R2}{R1(R2 + R3)} \quad \text{(approx.)}$$

The transistor collector current, and thus the load current is then the same:

$$\frac{V_{in} R2}{R1(R2 + R3)}$$

provided the transistor is not saturated. This is ensured if V_{in} does not exceed $V_{in} R3/(R2+ R3)$.

Circuit values are calculated on the basis of the available supply voltage V_{in} and the maximum load voltage $V_{L(max)}$ for which the load current is to stay constant. Resistor R1 value should be high to make the current less dependent on transistor characteristics. Resistor values R2 and R3 are thus chosen to make the base voltage V 2 somewhat larger than $V_{L(max)}$ — say about 50 percent greater. At the same time normal transistor bias requirements must be met; also the base current multiplied by R2/R3 must produce negligible voltage drop compared with the voltage V1 across R2. If not the actual load current will be less than the required value. Also the load current will be more influenced by variations in transistor characteristics.

Almost any small signal pnp transistor can be used in this circuit (or an npn transistor with circuit polarity reversed), provided

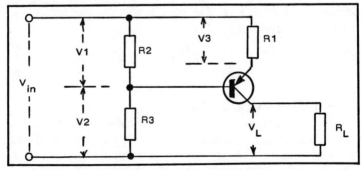

Fig. 30-2. Constant current from low voltage supply using a zener diode.

that its minimum value of β is 25 or more at 1 mA and it has a V_{ce} rating of at least 8 V with a base circuit resistance of 1.3 k ohm.

DESIGN EXAMPLE

Given source voltage available is 20 V. The load is resistive with a possible variation from 0 to 5 k ohms. A constant current output of 1 mA is required with a tolerance of plus or minus 5 percent (10 percent maximum swing).

(i) Calculate V_{Lmax} from known data

$$V_{Lmax} = \text{design current} \times \text{max load resistance}$$
$$= 0.001 \times 5000$$
$$= 5 \text{ V}$$

(ii) Choose a transistor with a β min of 25. Then maximum base current will be

$$\frac{\text{load current}}{25}$$
$$= \frac{1}{25}$$
$$= 0.04 \text{ mA}$$

(iii) Determine suitable values for R1, R2 and R3 from previous descriptions:

$$\frac{R3}{(R2 + R3)} = \frac{\text{voltage across R3}}{\text{voltage across (R2 + R3)}}$$

Since $V_{Lmax} = 5$, make voltage across R3 = 7.5 V
Voltage across (R2 + R3) is the supply voltage = 20

$$\text{So} \quad \frac{R3}{(R2 + R3)} = \frac{7.5}{20}$$

At this point we can determine a value for R1. We have to drop the supply voltage from 20 V to 7.5 V = 12.5 V across R1 with an emitter-base current of 1 mA, so:

$$R1 = \frac{12.5}{.001}$$
$$= 12.5 \text{ k ohms}$$

(nearest preferred value is 12 k ohms)

The base current can be allowed to 'swing' to the extent of plus or minus 5 percent. Its normal value (.04 mA) associated with load resistance value and $V_{L(max)}$ gives the equivalent of a voltage 'swing' across R2 of 0.06 V, so we can establish a basic relationship:

$$R2 \times 7.5/20 \quad \leq 0.06 \times 25 \text{ k ohms}$$
$$\text{or } R2 \quad \leq 4 \text{ k ohms}$$

and R3 = 7.5/voltage across R2

$$\text{Voltage across } R2 = 20 - 7.5$$
$$= 12.5 \text{ V}$$
$$\text{So } R3/R2 = 7.5/12.5$$
$$= 0.6$$

We now have to play with 'possible' values for R3 and R2, using preferred values and aiming to get as near as possible to R3/R2 = 0.6. The nearer we can get to this the better, otherwise this will modify the voltage across R2, which in turn will affect the design value of R1 and/or affect the current swing.

Probably R2 = 3.3 k ohms and R3 = 2.2 k ohms is about as near as we get, giving R3/R2 = 0.666. This should be reasonably satisfactory, bearing in mind that R1 has been adjusted to a preferred rule, and voltage across R2 has been estimated.

Check out the circuit for working with these values. If not satisfactory, adjust values of R1 and R2 (or R3) up or down a step or two, or recalculate values from first principles.

AN IMPROVED CIRCUIT

The improved circuit shown in Fig. 30-3 is basically the same as Fig. 30-2, but with R2 replaced by a zener diode. It is improved in the sense that whereas the original circuit depends on a constant voltage supply for output stability, this circuit will compensate for variations in supply voltage. The voltage across R1 is now almost equal to the zener voltage V_z, and constant within the practical limitations of a zener, provided V_z is substantially larger than the base-emitter voltage of the transistor. In other words, load current will be equal to $V_z/R1$, which is independent of the supply voltage V_{in}.

In practice the series resistance R_z of the zener diode has to be taken into account, which is typically of the order of 20 ohms for a small zener diode with a normal 10 - 12 zener voltage at a current of 5 mA.

The stability factor of the first circuit (Fig. 30-2) is given by:

$$\frac{(R2 + R3)}{R2}$$

The stability factor of the diode circuit (Fig. 30-3) is given by:

$$\frac{(R_z + R3)}{R_z}$$

The improvement offered by this second circuit is given by dividing the second by the first:

$$\frac{(R_z + R3) \times R2}{(R2 + R3) \times R_z}$$

Assuming the same values for R3 (say 2.2 k ohms), with R2 = 3.3 k ohms and taking R_z = 20 ohms as typical;

$$\text{improvement} = \frac{(10 + 2.200) \times 3300}{(2200 + 3300) \times 20}$$
$$= 66.3$$

In other words the zener circuit offers 66 times the stability of the previous circuit under conditions where the supply voltage is variable. In practice it would probably be greater than this since a lower value can be used for R3. Also, in this circuit the 'equivalent' resistance (i.e., the zener diode resistance) replacing R2 is now very much lower (20 ohms compared with 3,300 ohms), which reduces the effects of base current variations. There is thus justification for using this circuit even where the supply voltage is substantially constant.

OP-AMP CIRCUITS

Op-amp circuits are described elsewhere but specifically simple op-amp current stabilizing circuits have distinct limitations for use when the input voltage is variable (and at higher frequencies, as mentioned previously). A better circuit is shown in Fig. 30-4 where

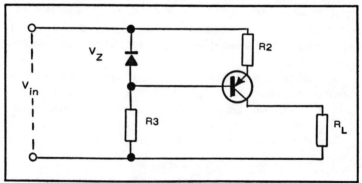

Fig. 30-3. Improved circuit based on zener diode.

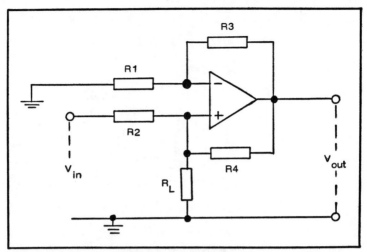

Fig. 30-4. Current stabilized circuit using op amp.

both input and output are referred to ground. This circuit is shown in the *inverting* mode. It can equally well be worked in the *noninverting* mode by feeding V_{in} through R1 to the negative input terminal of the op amp and grounding the positive input terminal through R2.

In a practical circuit of this type R1 = R2, and R3 = R4. The load current for any rate of load resistance R_L is then given by:

$$\text{load current} = \frac{-V_{in}}{R1}$$

This is limited in practice by the magnitudes that the load current and V_{out} could reach. If V_{out} reaches the amplifier limit the feedback is broken and the simple current relationship no longer applies.

Chapter 31

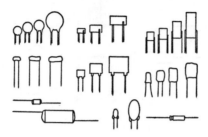

Bootstrapping

It is a common requirement that a high input resistance is required in a transistor stage (e.g., in an amplifier stage). In a straightforward circuit input resistance is provided by the base resistor forming part of the bias circuit (with common emitter configuration). However to minimize base circuit variations in order to avoid excessive temperature drift the maximum value which can be used for the base resistor is limited. In practice, values of up to 500 k ohm may be used with silicon transistors, but generally less than 10 k ohms in the case of germanium transistors.

The basic rule emerges—use a silicon transistor if the circuit demands a high input resistance. But this may still not be enough. A possible solution is then to apply 'bootstrapping'—a technique to make a lowish value base resistor present a very much higher value to the input signal. It is called 'bootstrapping' because the resistor acts as if it were "pulling itself up (in value) by its own bootstraps" with no apparent external help. In fact, it does get help in the form of a feedback provided by a capacitor connected between the emitter and the bias capacities R1 and R2, as shown in Fig. 31-1.

A large value is chosen for this capacitor so that it will act as a short circuit at the lowest frequency under consideration. Under this condition the bottom of R3 will effectively be connected directly to the emitter and output, while the top of R3 is connected to the input. The effective input resistance is then:

$$R \text{ effective} = \frac{R3}{1 - Av}$$

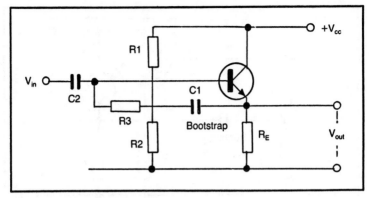

Fig. 31-1. Basic bootstrap circuit.

where Av is the voltage gain

The value of Av, in fact, approaches 1. Hence with this particular biasing arrangement the value of the effective input resistance becomes very high indeed. In practice Av will never reach a value of 1 - possibly 6.995, say. Then, if R3 is 100 k ohms, say, the effective input resistance could be:

$$\frac{100\ k}{1 - .995} = 200\ M\ ohms$$

At the same time R3 appears large to the transistor base current although the effect on its working properties will be negligible (i.e., will have negligible effect on the 'working' values of R1, R2 and R3).

The chief difficulty in the design of a bootstrapped circuit is deciding on a suitable value for the capacitor C1. Logically this should be excessively large to ensure bootstrapping or low frequencies (the larger the better in this respect). Unfortunately this can

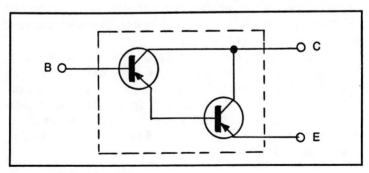

Fig. 31-2. Darlington transistor pair.

261

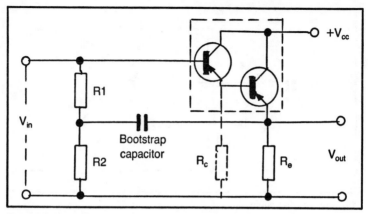

Fig. 31-3. Bootstrap circuit based on Darlington pair.

upset low frequency gain, particularly in a capacity coupled stage. Thus ideally it should be made no larger than necessary.

Consider first the effect of the values of R3 and R2 on the bootstrap performance. For dc bias purposes the base resistor is R3 + R2, but only R3 becomes bootstrapped. If R3 is made much smaller than R2, then the bootstrapped value of R3 may not be large enough. Equally, a small value of R2 will reduce input resistance. The practical compromise is to make R3 between ¼ and ¾ of R2.

A 'possible' value for C1 is then 1 μF. Try adjusting it up or down in a practical circuit. Another method of adjusting the final circuit is to add a capacitor and parallel resistor in series with the bootstrap circuit (i.e., between C1 and R3) to reduce feedback at low frequencies.

Alternative to bootstrapping a Darlington transistor pair can be used to produce a high input impedance. This circuit consists of two transistors forming a composite pair, the input resistance of the second transistor forming the emitter resistor (load) for the first transistor. This device is commonly available in a single (transistor) package with three leads—Fig. 31-2.

Again a Darlington pair can be bootstrapped, if necessary, using the circuit shown in Fig. 31-3. In this case R1 is the equivalent of R3 in the primary description, with R2 the same as R2. The input resistance in this case is substantially equal to the product of the short circuit current gain and the effective emitter resistance R_e. Very high current gains can be provided by Darlington pairs—e.g., from 50 up to 30,000.

Chapter 32

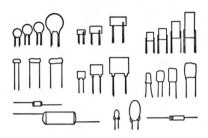

Logic Gates

Circuits working on *logic* employ a system of *gates* which can be constructed from discrete components or, now almost universely, employing integrated circuits. A single IC, for example, may contain a considerable number of individual gates in a single package.

The vast majority of logic elements work on the binary number system, i.e., 'counting' in only two digits (known as *bits*) - 0 (zero) and 1 (one). Equally, this corresponds to a basic switching action, 'off' or 'on', respectively. Logic elements are thus *digital* devices. They do not necessarily work on 0 or 1 (off or on) basis. More conversely they are based on the difference between two *dc* voltage levels. If the more positive voltage signifies 1, then the system employs *positive* logic. If the more negative voltage signifies 1, then the system employs *negative* logic. It should be noted that in both cases, although the lower or higher voltage respectively signifies 0, this is not necessarily a *zero* voltage level, so the actual voltage values have no real significance.

There is another system, known as *pulse-logic*, where a 'bit' is recognized by the presence or absence of a pulse (positive pulse in the case of a positive-logic system and negative pulse in the case of a negative-logic system).

GATES

Logic functions are performed by a number of logic *gates*. The three basic logic functions are OR, AND and NOT. All are designed to accept two or more *input* signals and have a single *output* lead.

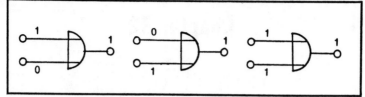

Fig. 32-1. Logic OR gate.

The presence of a signal is signalled by 1 and the absence of a signal by 0.

The four possible states of an OR gate with two inputs (A and B) are shown in Fig. 32-1. There is an *output* signal whenever there is an input signal applied to input A OR input B (and also with input at A and B simultaneously). This applies regardless of the actual number of inputs the gate is designed to accept. The behavior of an OR gate (again written for only two inputs) is expressed by the following *truth table:*

A	B	output (Y)
0	0	0
0	1	1
1	0	1
1	1	1

It can also be expressed in terms of Boolean algebra, calling the output Y

$$Y = A + B + \ldots + N$$

where N is the number of gates

The important thing to remember is that in Boolean algebra the sign + DOES NOT MEAN PLUS' BUT OR.

The AND gate again has two or more inputs and one output, but this time the output is 1 only if *all* the inputs are also 1. The truth table in this case is quite different - Fig. 32-2. The corresponding equation of an AND gate is:

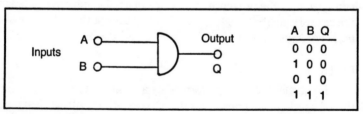

Fig. 32-2. Logic AND gate.

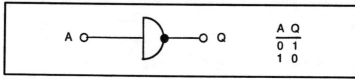

Fig. 32-3. Logic NOT gate.

$$Y = A \cdot B \ldots N$$
$$\text{or, } Y = A \times B \ldots \times N$$

This time the • or × sign does not mean 'multiplied by' as in conventional arithmetic, but AND.

The NOT gate has a single input and a single output - Fig. 32-3, without output always opposite to the input, i.e. if $A = 1$, $Y = 0$ and if $A = 0$, $Y = 1$. In other words it inverts the sense of the output with respect to the input and is thus commonly called an *inverter*. Its Boolean equation is:

$$Y = \overline{A}$$
(Y equals NOT A)

Combinations of a NOT gate with an OR gate or AND gate produce a NOR and NAND gate, respectively, working in the inverse sense to OR and AND.

Diode-logic (DL) circuits for an OR gate and an AND gate are shown in Fig. 32-4. Both are shown for negative logic and are identical except for the polarity of the diodes. In fact a positive-logic DL or OR gate becomes a negative-logic AND gate; and a positive-logic AND gate a negative-logic OR gate.

The simple NOT gate or inverter shown in Fig. 32-5 is based on a transistor logic—an npn transistor for positive-logic and a pnp transistor for negative-logic. The capacitor across the input resistance is added to improve the transient response.

PRACTICAL GATES

Most logic gates are produced in the form of integrated circuits, from which various 'family' names are derived. NAND and

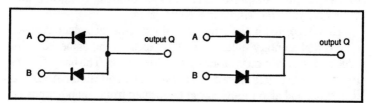

Fig. 32-4. Diode logic circuit for OR gate (left) and AND gate (right).

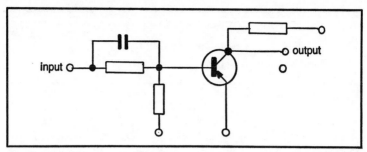

Fig. 32-5. NOT gate or Inverter.

NOR gates, for example, are a combination of AND or OR gates, respectively, with a NOT gate inverter. From the basic circuits just described, such functions can be performed by diode-transistor logic or DTL gates.

Faster and rather better performance can be realized with transistor-transistor-logic gates (TTL). During the early 1970s DTL and TTL represented the bulk of the IC digital productions, but since then various other IC families have appeared, each offering specific advantages and more functions for particular applications. These are:

☐ RTL (resistor-transistor logic) which can be made very small-even by microelectronic standards - and is capable of performing a large number of functions.

☐ DCTL (direct-coupled-transistor logic), which employs the same type of circuit as RTL but with the base resistors omitted. This gate, which can perform NOR or NAND functions, has the advantage of needing only one low voltage supply and has low-power classification.

☐ HTL (high threshold logic) is based on diode-transistor logic similar to DTL but also incorporates a Zener diode to stabilize the circuit and provide high immunity to 'noise'. It is usually chosen for applications where this feature is important. MOS (metal oxide semiconductor logic), based entirely on field effect transistors (FETs) to the complete exclusion of diodes, resistors and capacitors, yielding NAND and NOR gates.

☐ CMOS (complimentary metal-oxide-semiconductor logic) using complimentary enhancement devices on the same IC chip, reducing the power dissipation to very low levels. The basic CMOS circuit is a NOT gate (inverter), but more complicated NAND and NOR gates and also flip-flops can be formed from combinations or smaller circuits (again in a single chip).

☐ ECL (emitter-coupled logic) also known as CML (current-mode logic). This family is based on a differential amplifier which is basically an analog device. Nevertheless it has important application in digital logic and is the faster operating of all the logic families with delay times as low as 1 nanosecond per gate.

FAN-IN AND FAN-OUT

The terms *fan-in* and *fan-out* are used with IC logic devices. Fan-in refers to the number of separate inputs to a logic gate. Fan-out is the number of circuit loads the output can accommodate, or in other words the number of separate outputs provided. Fan-out is commonly 10, meaning that the output of the gate can be connected to 10 standard inputs on matching gates. Each separate input represents a load, the higher the number of separate loads the higher the current output of the device providing fan-out needs to be in order to provide the *standard load* on each input, i.e. passing enough current to drop each input voltage to the design figure.

It is possible to increase fan-out by replacing diode(s) with transistor(s) in the device concerned, so 10 is by no means a maximum number.

FLIP-FLOPS

A flip-flop is a bistable circuit and another important element in digital logic. Since it is capable of storing *one* bit of information it is functionally a *1-bit memory unit.* Because this information is locked or 'latched' in place, a flip-flop is also known as a *latch.* A combination of n flip-flops can thus store an *n-bit word,* such a unit being referred to as a *register.*

A basic flip-flop circuit is formed by cross-coupling two single-input NOT gates, the output of each gate being connected back to the input of the other gate - Fig. 32-6. However, to be able to

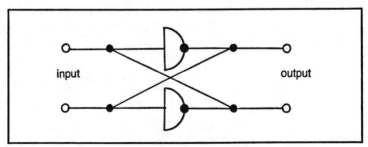

Fig. 32-6. Basic flip-flop circuit formed by cross-coupling two NOT gates.

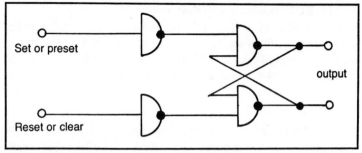

Fig. 32-7. Extension of the same principle to provide preset or clear, using two-input NOT gates.

preset or clear the state of the flip-flop, two two-input NOT gates cross-coupled are necessary. Each is preceded by single-input NOT gates cross-coupled are necessary, each preceded by single-input NOT gates as shown in Fig. 32-7. When the flip-flop is used in a pulsed or clocked system the preceding gates are known as the *steering* gates with the cross-coupled two-input gates forming the *latch*. This particular configuration is also known as a S-R or R-S flip-flop.

Two other variations of the flip-flop are also produced as integrated circuits: *J-K flip-flop* - which is an S-R flip-flop preceded by two AND gates. This configuration removes any ambiguity in the truth table. It can be used as a T-type flip-flop by connecting the J and K inputs together (see Fig. 32-8 for connections). *D-type flip-flop* - which is a J-K flip-flop modified by the addition of an inverter. It functions as a 1-bit delay device.

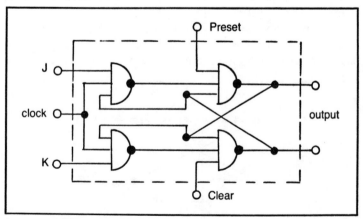

Fig. 32-8. J-K flip-flop.

DTL or TTL logic gates tend to have an intrinsic low value of *dc* noise immunity which, for more critical applications may demand accurate power supplies, filtering and decoupling networks, electrostatic and magnetic shields, etc. Other logic families (HTL, MOS, etc.) do not suffer from such limitations to any extent. Where these are not specifically referred to by name, they can be generally described as High level logic (HLL) devices.

The following is a description of a typical HLL family by SGS providing high level logic with DTL and TTL devices, which not only describes logic gates and their function in practical terms but also illustrates how the performance characteristics of such devices are presented, analyzed and applied. This particular HLL family includes the following devices:

- ☐ H 102 : quad two input NAND gate
- ☐ H 103 : triple three input NAND gate
- ☐ H 104 : dual four input NAND gate with expander inputs
- ☐ H 109 : dual four input AND power gate with expander inputs
- ☐ H 110 : dual J-K flip-flop with asynchronous set input
- ☐ H 111 : dual J-K flip-flop with asynchronous set and clear
- ☐ H 113 : quad high to low level converter or open output collector HLL NAND gate
- ☐ H 114 : quad low to high level converter
- ☐ H 122 : quadruple two input NAND gate with passive pull-up
- ☐ H 124 : dual four input NAND gate with passive pull-up
- ☐ H 157 : decade counter

The packaged forms of these devices together with pin-cut identification and internal connections are shown in Fig. 32-9.

THE BASIC GATE CIRCUIT

Figure 32-10 shows the basic gate circuit of the HLL family. It is evident that this circuit performs the NAND function for positive logic or NOR function for negative logic.

To explain the gate circuit more fully let us divide it into four sections. Section A is the logic input gate composed of common emitter pnp transistors. The logic function of section A is to perform the 'AND' operation. Pnp transistors were chosen in place of diodes because they add current gain.

Each input of the gate requires an input current of only 0.48 mA (typical), which explains the very high fan-out of the family.

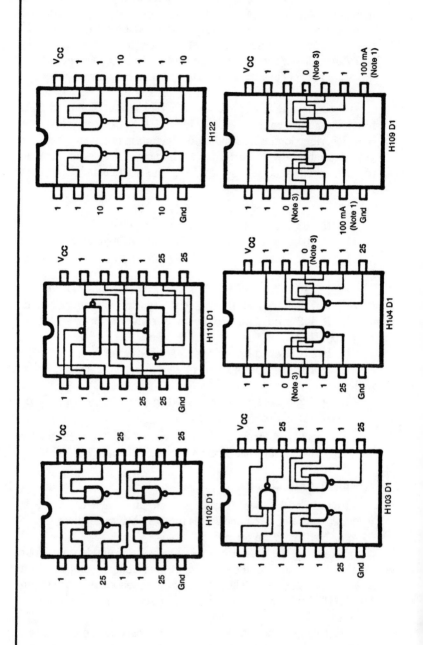

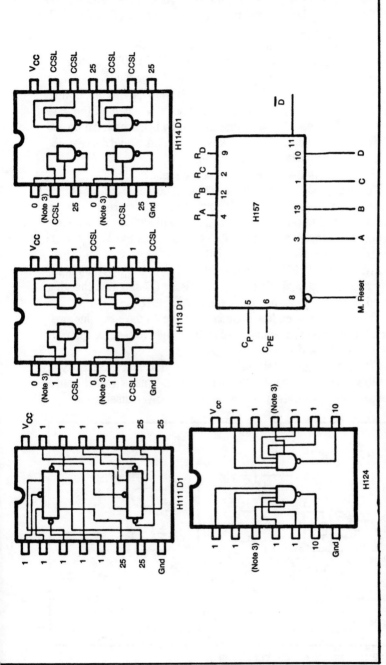

Fig. 32-9. Package forms of HLL family of logic gates.

271

Section B is composed of transistor Q5, and resistors R1 and R2. They form a buffer amplifier similar to that used in the DTL family.

Section C is the threshold setting Zener diode (6 V) which replaces the corresponding diode (0.6 V) of DTL. The high noise immunity of the HLL family is due to this zener diode.

Diode D1 serves to discharge point P rapidly; if this diode were not inserted, point P, which rises to about 6 V when all the gate inputs are high, would have to discharge only by recombination and leakage when one or more inputs went low. This would decrease *ac* noise immunity during discharge times.

Section D is the inverting output stage.

Transistor Q6 - Q7 and diode D2 form the active pull-up output circuit: when Q6 is 'ON' (i.e., all inputs are high) current from the load connected to point N is sunk through diode D2. As a consequence the output low voltage, V_{OL} is made up of the $V_{CE(sat)}$ of Q6 plus the forward voltage drop of D2.

When Q6 is 'OFF' (i.e., one or more inputs are low) the output point N goes high and the pull-up transistor Q7 supplies charging current to any capacitance connected at point N.

This also ensures low impedance drive to the load, and provides for ac noise immunity. The output high voltage is approximately one diode drop (0.8 V) below V_{CC}.

The advantages of this output configuration over the TTL type of pull-up circuit is the elimination of current 'spikes' (which cause internal noise generation and unnecessary power dissipation.

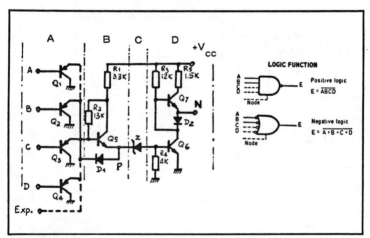

Fig. 32-10. Basic gate circuit of HLL family.

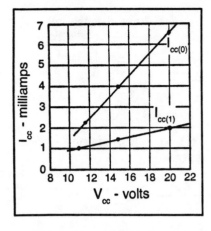

Fig. 32-11. Absorbed current as function of supply.

Recommended supply voltage for HLL elements is 15 V. Most of the characteristics described in this and in the following paragraphs apply at $V_{cc} = 15$ V.

Figure 32-11 shows the trend of the absorbed current of the circuit as a function of supply V_{cc} for the two logic stages 1 and 0. Similarly, Fig. 32-12 shows the average power absorption as a function of V_{cc}. In some cases parameter variation with supply voltages is given.

DC PARAMETERS

Dc parameters are measured using the following values of 'Forcing functions' (see later for definitions):

$$
\begin{array}{llll}
V_{CC} & = & 15\ V & \qquad V_{Fex} & = & 2\ V \\
V_{IL} & = & 6\ V & \qquad V_{R} & = & 20\ V \\
V_{IH} & = & 8\ V & \qquad I_{OH} & = & 0.2\ mA \\
V_{F} & = & 1.5\ V & \qquad I_{OL} & = & 12\ mA
\end{array}
$$

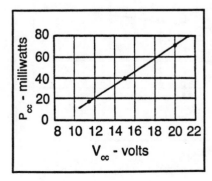

Fig. 32-12. Absorbed power as function of the supply.

Characteristics of the devices has shown that the following worst case parameter values can be guaranteed (at 25°C):

$$V_{OH} = \geq 13.5 \ V \qquad I_{Fex} = \leq 1.33 \ mA$$
$$V_{OL} = \leq 1.5 \ V \qquad I_{R} = \leq 5 \ \mu A$$
$$I_{F} = \leq 0.48 \ mA \qquad I_{SC} = \leq 25 \ mA$$

FAN-OUT

Fan-out is given by the ratio of I_{QL} to I_F. According to the values given above, fan-out in the worst case is:

$$FO = \frac{I_{OL}}{I_{F(max)}} = \frac{12 \ mA}{0.48 \ mA} = 25$$

Fan-out value = 25 is valid for V_{cc} min 10.8 V in the ambient temperature range 0°C − 75°C.

For increasing supply voltage values the fan-out increases as in Fig. 32-13. The increase is shown as relative to fan-out and fan-in at 10.8 V_{cc}.

The high FO capability is made possible by the use of input pnp transistors, in place of diodes. I_F, in fact, is the base current of these transistors (see Fig. 32-14), while their emitter current is essentially identical to the expander input current (1) I_{Fex}, which is typically 0.9 mA and 1.33 mA max. The minimum gain the pnp

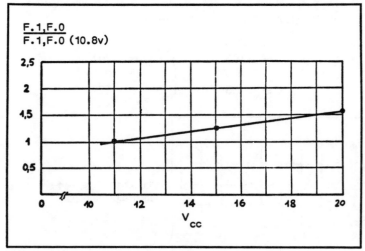

Fig. 32-13. Fan-out increase with voltage.

transistor should have is then approximately:

$$h_{fe} = \frac{I_{Fex}}{I_F} - 1 = \frac{1.33}{0.48} - 1 = 1.8$$

This current can be measured only when expander inputs are provided, i.e., in H104 and H109 etc.

In practice the pnp gain is higher, in the range of 4 or 5, correspondingly I_F is lower, and fan-out higher.

A different situation arises when diodes are connected to the expander inputs to increase the number of inputs to the gate. In this case, obviously, I_F coincides essentially with I_{Fex}. (1.33 mA worst case).

Therefore worst case fan-out of a gate with respect to the diode expanded input is:

$$FO = \frac{I_{OL}}{I_{Fex}} = \frac{12 \text{ mA}}{1.33 \text{ mA}} = 9$$

In general, when one gate drives a combination of standard inputs and of diode inputs, the following condition must be met:

$$S \cdot I_F + D \cdot I_{Fex} \leq I_{OL}$$

where S is the number of standard inputs and D the number of diode inputs connected to the gate. This formula allows the calculation of S or D, when one of them is known. Figure 32-15 indicates the typical variation of I_F with temperature, and may be used to have an indication of the temperature dependence of fan-out.

NOISE IMMUNITY

Worst case dc noise immunity can be calculated from the dc data above. These give:

Fig. 32-14. Transistor currents.

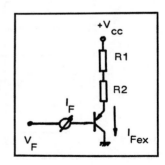

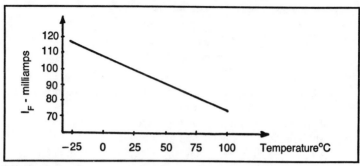

Fig. 32-15. Variations in I_F with temperature.

(i) high level dc noise immunity $= V_{OH} - V_{IH}$
$$13.5 - 8 = 5.5 \text{ V}$$

(ii) low level dc noise immunity $= V_{IL} - V_{OL}$
$$6 - 1.5 = 4.5 \text{ V}$$

Figure 32-16 shows a typical transfer characteristic of the gate: it is apparent that the true threshold of the gate lies almost exactly midway between V_{IL} (6 V) and V_{IH} (8 V).

It is interesting to note that the true threshold of the gate, and hence V_{IL} and V_{IH}, is essentially constant with temperature due to the almost exact compensation of the temperature coefficients of the threshold determining devices in the circuit. More specifically, the V_{BE} drop of the input pnp is temperature compensated by the V_{BE} drop of the input buffer transistor (T5 in Fig. 32-10); and the positive temperature coefficient of the zener diode (Z in Fig. 32-10) is compensated by the negative temperature coefficient of the V_{BE} of the output transistor (T6 in Fig. 32-10).

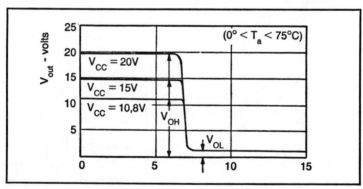

Fig. 32-16. Typical transfer characteristics of the gate.

Thus dc noise immunity changes very little with temperature; any changes depend essentially on the temperature coefficients of V_{OL} and V_{OH}. Since V_{OL} is the sum of one V_{BE} drop plus one $V_{CE(sat)}$ which have opposite temperature coefficients, there is compensation to some extent; V_{OH} is always one V_{BE} drop (0.8 V) below V_{CC}, and therefore it increases at a rate of about 2 mV/°C with temperature. As a conclusion, low level dc noise immunity is almost constant whereas high level dc noise immunity increases slightly with temperature. The dependence of dc noise immunity on supply voltage will now be examined.

The threshold of the device is essentially independent of V_{CC}, and the same applies to V_{OL}. Thus, of the noise immunity determining parameters, only V_{OH} is about one V_{BE} drop below V_{CC}. Therefore only the high level noise immunity is affected by V_{CC}.

For example we have at the min operating supply voltage $V_{CC} = 10.8$ V:

(i) high level noise immunity:

$$V_{OH} - V_{IH} = V_{CC} - 1.5 - V_{IH} = 10.8 - 1.5 - 8 = 1.3 \text{ V}$$

and at the max operating supply voltage, $V_{CC} = 20$ V.

(ii) high level noise immunity:

$$V_{CC} - 1.5 - V_{IH} = 20 - 1.5 - 8 = 10.5 \text{ V}$$

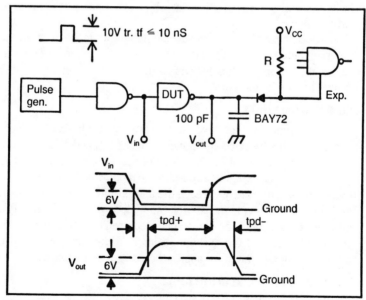

Fig. 32-17. Switching time test circuit and measurement of conditions.

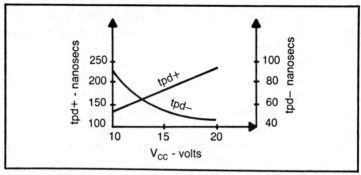

Fig. 32-18. Tpd variations with V_{CC}.

AC PARAMETERS

Figure 32-17 shows the switching time test circuit, and the measurement conditions. Resistor R, in the load circuit, is chosen so that the load current of the DUT is approximately 12 mA. A plot showing variation of tpd+ and tpd– with V_{CC} is given in Fig. 32-18.

The variation of propagation delays as a function of load capacitance and V_{CC} is given in Fig. 32-19.

The test circuit used in this measurement is shown in the insert to the diagram. Finally, Fig. 32-20 shows the variation of propagation delays as a function of fan-out and V_{CC}. This is particularly interesting for HLL devices, which are capable of very large fan-out.

The test used for this measurement is shown in the insert of the diagram; note that each fan-out corresponds to one input of a separate gate, so that the fan-out equals the number of separate gates driven by one single gate.

AC NOISE IMMUNITY

Low level and high level ac noise immunity have been measured, using the test circuits shown in Figs. 32-21 and 32-22 respectively. Ac noise immunity is here defined as the amplitude and duration of a voltage 'spike' applied at the input, which will cause the output of the gate barely to reach the appropriate threshold voltage (V_{IH} for the low level, and V_{IL} for the high level noise immunity). Obviously, the amplitude of the spike required to obtain this condition is a function of the duration of itself; if the duration tends to infinity, *ac* noise immunity approaches *dc* noise immunity.

THE POWER GATE CIRCUIT (H109)

Because of the intrinsically high fan-out capability of standard

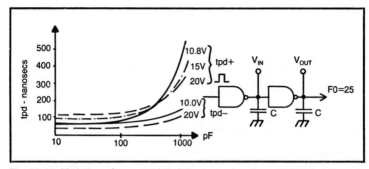

Fig. 32-19. Variation of propagation delays as a function of load capacitance.

HLL gates, a buffer-amplifier as conceived in other families of logic integrated circuits, is not needed in the HLL family.

Many industrial applications, however, require the use of a 'power' gate, i.e., a gate capable of switching relatively high currents, as, for example, in lamp and relay driving. The H109 power-gate was designed for this purpose. It consists of two four-input expandable AND gates, each corresponding to the circuit of Fig. 32-23.

It is, essentially, a circuit identical to that of the basic gate, with the addition of a Darlington inverter-power amplifier at the output. The final transistor is of a special design to handle the high current, typically in excess of 100 mA. At this current the value of V_{OL} is guaranteed to be less than 1.5 V, as for the standard gates. The output device does not have any pull-up circuitry; this allows output OR-ing. An external pull-up resistor can obviously be used if necessary. In normal applications, however, the load device (i.e.,a lamp, a relay) will take its place. The input characteristics of the power gate are identical to those of the basic gate.

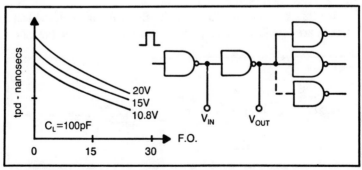

Fig. 32-20. Variations in propagation delays as a function of fan-out and V_{CC}.

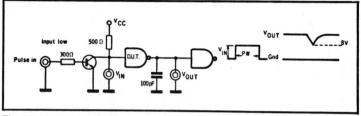

Fig. 32-21. Low-level noise immunity.

FLIP-FLOP (H110 - H111)

The J-K flip-flop configuration has been adopted in the HLL family. There are two flip-flop versions available; the first is a dual J-K flip-flop with only one asynchronous input per flip-flop in a 14 lead DIP package; the second is exactly identical to the first except that there are two asynchronous inputs per flip-flop in a 16 lead DIP package. The flip-flop circuit is shown in Fig. 32-24. Its operation will be explained with the help of the functional diagram in Fig. 32-25.

Boxes D1 and D2 represent circuits which generate voltage 'spikes' of limited duration when their inputs are enabled and the clock pulse undergoes a high to low transition. Inputs I1, I2 and I3 to these boxes are such that, when any one of them is low, spike generation is inhibited. Gates G1 and G2 form the basic memory element. The operation is then as follows:

Assume Q is high and $\overline{Q}$ low; if inputs J, K, C_D and S_D are all high, then spike generator D2 is enabled (all inputs are high), whereas D1 is disabled (because Q is low).

When C undergoes a negative transition, a negative voltage spike will appear at one of the inputs of G2, turning it off; as a consequence G1 will turn on. The same cycle will now be repeated, with Q and $\overline{Q}$ reversed.

In this situation, therefore, the flip-flop will toggle. If the J and K inputs are low; both spike generators are disabled and the clock

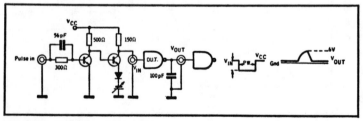

Fig. 32-22. High-level noise immunity.

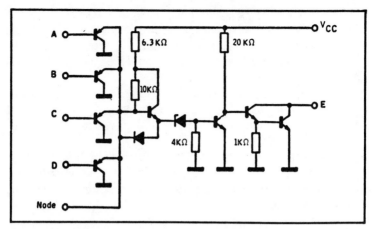

Fig. 32-23. H109 power gate design.

pulse negative transition will not affect the state of the flip-flop.

Finally, if J is high and K low or vice versa, this information will be transferred to the outputs upon arrival of the negative clock transition.

The operation of the asynchronous inputs C_D and S_D is also evident from the schematic in Fig. 32-25 when one of them is low, the output flip-flop will be modified to assure the desired state, and,

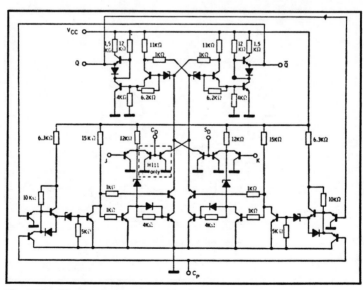

Fig. 32-24. Flip-flop circuit.

at the same time, spike generation is inhibited at the proper side; in this way the clock pulse has no effect on the state of the output flip-flop and no undesired spurious signals appear at the outputs. From a circuit point of view, the operation of the spike generators is as follows:

Assume Q, K, S_D and Cp are high; T1 and T3 are then 'ON' and T2 and T4 'OFF'. When Cp goes low T1 turns 'OFF', but its storage time and a controlled base discharge through resistor R cause the turn-off to be delayed.

During this storage period T3 and T4 are therefore 'ON' at the same time and a voltage spike appears at point M, which will cause $\overline{Q}$ to go high and Q to go low. If inputs K and/or S_D are low, T3 will be 'OFF' independently of Cp, thus inhibiting spike generation.

HIGH TO LOW LEVEL CONVERTER (H113)

In order to adapt the HLL family to low level families (DTL-TTL etc.) a high to low level converter has been designed. The circuit is as indicated in Fig. 32-26. It is composed of a normal gate except that the output transistor collector is open. The output voltage swing is therefore a function of voltage at which the pull-up resistor is connected.

Considering now the two possibilities of V_{OL} and V_{OH} we can consider the fan-out as indicated in Figs. 32-27 and 32-28.

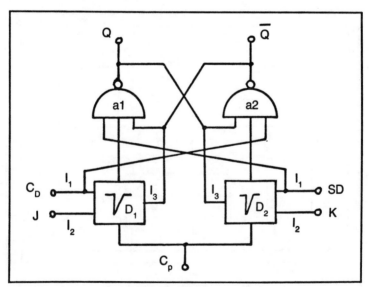

Fig. 32-25. Functional diagram for flip-flop circuit.

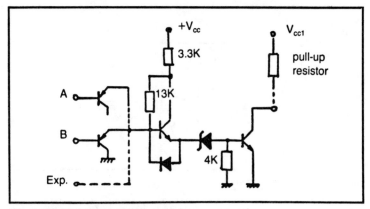

Fig. 32-26. High to low level converter.

LOW TO HIGH LEVEL CONVERTER (H114)

This gate is similar to the standard gate, except that the zener has been replaced by a diode, and the input resistors have been inverted in order to decrease the voltage across the diode. Figure 32-29 shows the electrical circuit of the H114.

FAN-IN EXTENSION

As already mentioned, fan-in can be easily increased by means of external diodes in the devices where an expander input is provided, as in the H104, H109, H113, and H114.

When an expander input is not available fan-in extension can also be obtained by connecting external diodes, and an auxiliary resistor R to V_{cc}, to any one of the standard inputs. This is shown in Fig. 32-30.

The value of resistor R should be determined according to the desired speed of the circuit; in fact any spurious capacitance present

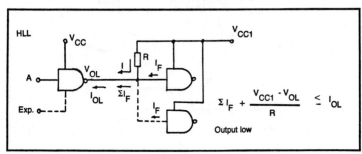

$$\Sigma I_F + \frac{V_{CC1} - V_{OL}}{R} \lesssim I_{OL}$$

Fig. 32-27. Low output fanout.

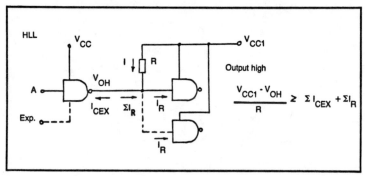

Fig. 32-28. High output fan-out.

at point A will cause propagation delay to increase (more precisely, it will increase tpd− in inverting gates and tpd+ in non-inverting gates).

For a given capacitance at point A, the appropriate propagation delay will increase as the values of R are increased.

When using this method of fan-in extension, care must be taken that they do not exceed the driving capability of the gates connected to the extended inputs; in fact the I_F at the standard input and of the current supplied through R. Another effect of this fan-in extension circuit is a slight reduction in low-level noise immunity. It is evident, in fact, that the V_{IL} of the circuit will now be decreased by one diode drop V_D. Therefore low level noise immunity will be, approximately:

$$V_{IL} - V_D - V_{OL} = 6 - 0.8 - 1.5 = 3.7 \text{ V}$$

in place of the standard 4.5 V mentioned in paragraph 4.1.

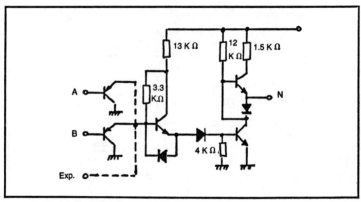

Fig. 32-29. H114 low to high level converter.

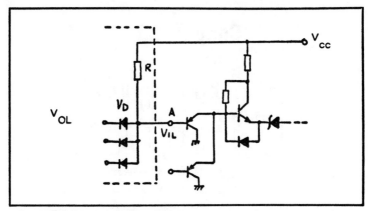

Fig. 32-30. Extension of fan-out through external diodes.

INCREASING PROPAGATION DELAYS

In those applications requiring low operation speeds it may be desirable to increase propagation delays in order to make the circuit less sensitive to ac noise. This may be achieved easily by connecting small capacitors at strategic points in the circuits. If the circuit is a gate with an expander input available to the outside, a capacitor can be connected to this point. This will increase either tpd− or tpd+ depending on whether the gate is inverting or non-inverting.

Definition of Symbols

V_{CC}	supply voltage
V_{IL}	input 'low' voltage
V_{IH}	input 'high' voltage
V_F	input voltage at which I_F is measured
V_{FEX}	input voltage at which I_{FEX} is measured
V_R	input voltage at which I_R is measured
I_{OH}	current drawn from output during V_{OL} measurement
V_{OH}	output 'high' voltage
V_{OL}	output 'low' voltage
I_F	input forward current
I_R	input reverse current
I_{FEX}	input current at the expander input
I_{SC}	output short circuit current

Chapter 33

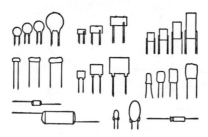

Hi-Fi

No designer's reference book would be complete without some mention of Hi-Fi although this is a complete, highly specialized subject on its own. Hi-Fi means simply high fidelity reproduction, although this is difficult to define exactly—it can be done by objective standards but assessment of what is 'high fidelity' reproduction is subjective and can vary widely with different individuals.

In practical terms, Hi-Fi means that the full audible frequency range should be reproducible—i.e., 20-30 Hz to 16-20,000 Hz, with minimum distortion. However, the ability to hear frequencies above about 12,000 Hz varies from individual to individual, and also deteriorates with age. The main difference between 'Hi-Fi' and ordinary 'domestic' sound reproduction lies at the extremes of the frequency range—e.g., below 100 Hz or reproduction of true bass sounds; and above 10,000 Hz, or true reproduction of treble. By comparison the ordinary domestic radio may have a frequency response of 100 Hz to 5,000 Hz.

Dynamic range is also significant, being the difference between the softest and loudest sounds in a presentation of music. The dynamic range of an orchestral work may be as high as 60-70 dB, although in actual recording this may be compressed to 60 dB or less.

The ability to cover dynamic range in reproduction is expressed by the signal-to-noise ratio, although this is varying with frequency. The most reliable signal-to-noise figures are thus those measured with 'weighting' (A-level weighting) to correspond to the typical responses of the human listener to the relative loudness of sounds of different frequencies.

Other parameters of significance are:

Distortion - which again will vary with both volume and frequency. The standard for measurement correctly taken is the measured distortion on the third harmonic of a 315 Hz tone recorded at response level. The average individual may not be able to detect distortion of less than 5 percent, although the 'true' ear can detect distortion of 2 percent or less.

Wow and Flutter - which are pitch changes produced by momentary departures from constant speed during playback. 'Wow' is the result of a brief loss of speed stability, resulting in a 'wobbling' of sustained notes. 'Flutter' is of imposing a high frequency 'flutter' on notes. Wow and flutter performance is usually expressed as a single measurement of loss of speed stability expressed as a percentage, but numerical values are dependent on the method of measurement.

Standard speeds are:

☐ Discs - 33 rpm or 45 rpm
☐ Tapes and Cartridges - 3¼ inches per second
☐ Cassette tapes - 1⅞ inches per second

Other significant performance parameters are detailed in the following extract from the DIN recommendations for Hi-Fi Standards in playback equipment.

Din Hi-Fi Standards (DIN 45 500)
(Minimum Performance Requirements)
Record Player Requirements

Speed tolerance	+1.5 percent −1 percent
Wow and flutter (peak)	± 0.2 percent
Rumble	−35 dB (unweighted)
(ref 1 kHz at 10 cm/sec)	−55 dB (weighted)
Pick-up	
Frequency response	40–12,500 Hz ± 5 dB
Channel balance	within 2 dB
Intermodulation distortion	1 percent
Crosstalk	
at 1,000 Hz	−20 dB
at 500-6,300 Hz	−15 dB
Max. playing weight	5g.
Compliance	at least 4×10^{-6} cm/dyne
Stylus tip radius	
spherical	0.6 + 0.1 thou
elliptical	0.78 × 0.24 thou

Stylus tip mass	2 mg.
Vertical tracking angle	15° ± 15°
Sensitivity crystal/ceramic magnetic	0.5-1.5V, 470k ohms 8-20mV, 47k ohms
Frequency response "flat" inputs equalized inputs	40-16,000 Hz ± 1.5 dB 40-16,000 Hz ± 2 dB
Channel balance	within 3 dB
Harmonic distortion pre-amplifier power amplifier	1 percent from 40-4,000 Hz 1 percent from 40-12,500 Hz and down to −20 dB
Intermodulation distortion Crosstalk (interchannel) at 1,000 Hz	−50 dB
from 250-10,000 Hz	−30 dB
Crosstalk (between inputs) at 1,000 Hz	−50 dB
from 250-10,000 Hz	−40 dB
Signal-to-noise ratio	50 dB
Output power mono amplifier	10 Watts
stereo amplifier	1 × 6 Watts
Inputs linear	500mV at 470K
magnetic pickup	5 mV at 47 k
Outputs pre-amplifier	1 volt at 47 k
to tape recorder	0.1 to 2 mV for each 1,000 ohms
speaker impedance	2, 4, 8, 16, 32, 50, 100, 400 or 800 ohms
damping factor	at least 3

Tape recorder requirements

Speed stability	± 1 percent over 30 seconds

288

Wow and flutter (peak)	± 0.2 percent
Frequency response	40-12,5000 Hz
Distortion for full	5 percent
modulation at 333 Hz	45 dB (unweighted)
Signal-to-noise ratio	50 dB (weighted)
Crosstalk (at 1,000 Hz)	
mono	−60 dB
stereo	−25 dB
Erasure	−60 dB

Loudspeaker requirements

Frequency response (axis)	50-12,500 Hz
Matching of stereo pairs	within 3 dB 250-8,000 Hz within 4 dB at 15° from axis up to 8,000 Hz
Polar response Sound pressure at 1 meter	12 microbars
at 3 meters	4 microbars
Distortion factor 250 to 1,000 Hz 1,000 to 2,000 Hz	3 percent falling from 3 percent to 1 percent
above 2,000 Hz	1 percent
Transient performance	Slope not to exceed 12 dB/octave in range 50-250 Hz
Impedance variation (over frequency range)	within 20 percent of nominal
Power handling capacity	10 Watts
Nominal impedance	4, 8, or 16 ohms

Vhf tuner requirements

Frequency response	40-12,500 Hz ± 3 dB 50-6,300 Hz ± 1.5 dB
Channel balance	within 3 dB 250-6,300 Hz

Harmonic distortion	2 percent for 40 kHz deviation
Crosstalk 250-6,300 Hz	26 dB
6,300-12,500 Hz	15 dB
Signal-to-noise ratio	54 dB
Pilot tone suppression at 19 kHz	20 dB
at 38 kHz	30 dB
Audio output	0.5 to 2 volts into 470 k

Requirements for integrated systems

Record player/amplifier	40-12,500 Hz ± 6.5 dB
Frequency response	63.5-8,000 Hz ± 3.4 dB
Channel balance	within 5 dB
Crosstalk at 1,000 Hz	−19 dB
from 500-6,300 Hz	−14 dB
Tape recorder/amplifier Crosstalk (stereo) at 1,000 Hz	−24 dB
from 250-10,000 Hz	−21 dB
Signal-to-noise ratio	41 dB
VHF tuner/amplifier Frequency/response	40-12,500 Hz ± 4.5 dB 50-6,300 Hz ± 3 dB
Channel balance (250-5,300 Hz)	within 6 dB
Harmonic distortion	2.5 percent
(40 kHz deviation)	
Crosstalk at 1,000 Hz	−24 dB
from 250-6,300 Hz	− 18 dB
from 6,300-10,000 Hz	−14 dB

Signal-to-noise ratio	50 dB
Pilot tone suppression at 19 kHz	−19 dB
at 38 kHz	−29 dB

HI-FI IN THE HOME

The final items which will govern your enjoyment of Hi-Fi are the performance of the loudspeakers (but particularly *your* assessment of their performance, not just the quality of the technical specification); and the size and shape of the room when you have installed them.

It is a characteristic of sound waves distributed in a room or similar enclosure that, apart from room acoustics, the listening quality may be considerably modified by the formation of standing waves, caused by resonance. This will result in a change of sound level and quality in different parts of the room.

Resonance will occur—with consequent formation of a standing wave—where one dimension of the room is equal to one half the wavelength of a particular sound being reproduced. In general this effect can be reduced, or eliminated, by altering the positioning of the speakers relative to the sound reflecting surfaces present in the room (mainly the walls and ceiling).

The same principle, however, establishes that the lowest frequency which can be reproduced in any room is governed by the room diagonal measurement (from one floor corner to the opposite corner of the ceiling). Frequencies with a wavelength greater than twice this diagonal measurement cannot be reproduced, i.e., will be cut off. The following information can be used as a specific guide to cut-off frequencies.

Cut-off frequencies

Cut-off frequency (Hz)	length of room diagonal* feet
60	9.25
50	11.2
45	12.4
40	14.0
35	15.9
30	18.5
25	22.4
20	28.0

*Length of room diagonal $= L^2 + W^2 + H^2$

where L = length of room
W = width of room
H = height of room

Example: Find the cut-off frequency for a room size 11 ft long by 10 ft wide by 8 ft high.

Length of room diagonal $= (11 + 11) + (10 \times 10) + 8 \times 8)$ $= 285 = 16.9$ ft.

Therefore cut-off frequency will be between 30 Hz (18.5 ft) say about 33 Hz.

In other words, even if your hearing is very acute you would not hear frequencies below about 33 Hz, even if your equipment was capable of reproducing them!

Chapter 34

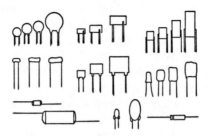

Hi-Fi Facts and Figures

This chapter is designed to take the 'mystery' out of reading, understanding and comparing manufacturers quoted performance and specification figures, and relating them to Hi-Fi standards. Actually, there is only one 'official' Hi-Fi specification which is widely accepted—called DIN 45 500 (manufacturers' literature will frequently refer to this standard). The figures quoted in this standard are regarded as *minimum* values for Hi-Fi reproduction. Meeting such figures in every respect qualifies equipment to be rated as true Hi-Fi, but much modern Hi-Fi equipment aims at bettering the DIN figures.

Remember, too, that the performance of each unit in the Hi-Fi system (or each *section* in integral systems) must be examined separately. The ultimate performance of any system cannot be better than that of the weakest unit (or lowest performance unit) in the chain—and may well be down on that due to mismatching!

Before attempting to read specification figures, however, it is necessary to understand what a *decibel* is. A decibel, written as dB, is a *ratio* between two sound levels, not a simple arithmetic measure. In a specification figure a ± dB value means a deviation above and below any frequency quoted—or more usually to a single *reference level* (333 or 315 Hz, although the actual figure may not be quoted).

Most people can detect a difference in sound levels of 3 dB, and the trained ear can detect a difference as low as 1 dB. *Frequency* deviations greater than + or − 3 dB would detract from Hi-Fi performance.

The *decibel* is also a direct measure of the loudness of a sound, a difference of + 6 dB representing a *doubling* of the original sound; and −6dB a halving of the original (or reference) sound. Thus the range of *tone controls*, etc., may be quoted in + or − dB.

Frequency Range & Frequency Response. The significance of frequency range has already been explained in the opening chapter. It is *the* basic requirement for high fidelity reproduction (the rest of the parameters are largely covered with noise 'content', distortion, etc.).

To meet DIN Hi-Fi requirements, minimum frequency ranges for various units are:

> *VHF Tuner*—40 - 12,500 Hz ± 3 dB
> *Amplifier*—40 - 16,000 Hz ± 1.5 dB
> *Loudspeaker(s)*—50 - 12,500 Hz
> *Record Players*—40 - 12,500 Hz ± 5 dB
> (pick-up)—63.5 - 8,000 Hz ± 2 dB
> *Tape Decks*—40 - 12,500 Hz

There is a subtle difference between frequency *range* (or frequency coverage) and frequency *response*. The former merely specifies the range of frequencies covered—e.g., 40-12,500 Hz. Frequency response indicates both the frequency range and the deviation from true or linear response over this range—e.g., 40-12,500 Hz ± 2 dB. However, this still does not show where—and how—this deviation occurs. Only a graph will do this, when the significant fact is how flat or near parallel the frequency response curve is over the main part of its range. It is better to have the deviation at the end(s) rather than in the middle of the graph.

SIGNAL-TO-NOISE RATIO

Signal-to-noise ratio, or Signal/Noise ratio (or abbreviated as S/N) is a measure of the difference between the softest and loudest sounds reproduced, or 'loudness difference' possible. Technically, this is called *dynamic range,* expressed in decibels. Thus a dynamic range of 60 dB represents a 'loudness difference' of 10.

Adequate *dynamic range* is necessary in order to hear the whole content of music. The dynamic range of a full orchestra may approach, or even exceed, 60 dB. That of 'pop' music is very much lower—perhaps only 25-30 dB (note that it is not the 'loudness' of sound that counts here, but the range of sounds from the softest to the loudest). So really you need a dynamic range of 60 dB or better fully to appreciate orchestral music playback—but a very much lower figure will be quite adequate for listening to 'pop'.

On tape machines, dynamic range may also be quoted as an *erase ratio,* measuring the dynamic range content of the tape when run for re-recording (also a measure of the performance of the erase head).

The S/N ratio on manufacturers specifications, given as a dB value, may also have added 'weighted', A-scale, or have dBA instead of just dB. This simply means that the measurement has been taken on a sound level meter compensated or 'weighted' to respond to different frequencies in a similar manner to the human ear. (The ear tends to assess higher frequency sounds as being 'louder' than they actually are.) A 'weighted' measurement is thus more realistic than an unweighted one.

Remember that a high S/N ratio will not *increase* the dynamic range of any particular recording. Records are usually best in this respect and a good disc may have a dynamic range as high as 65-70 dB (although 55-65 dB is more likely). Reel-to-reel tapes may have a dynamic range of 60-65 dB; and cassette tapes rather less. Again, manufacturers quote S/N ratios (dynamic range) for particular types of tapes.

DISTORTION

Overall DIN requirements are for a total harmonic (or 'musical') distortion of less than 5 percent, so any distortion figure quoted under 5 percent should be Hi-Fi, but the keen ear could detect distortion as low as 2 percent.

Unfortunately, a simple distortion figure quoted does not necessarily mean a great deal. Distortion tends to increase with the level (or loudness) of the sound, so one simple way of getting a low distortion figure would be to turn the volume control down when actually measuring distortion. Distortion level also varies with the frequency of the sound—again measured distortion figures being capable of 'adjustment' via the tone control!

There is, however, a standard for measurement of distortion, based on the playback of a specific frequency (315 Hz) at a specified sound level (0 on the VU or volume level on decks). 'Standard level', or similar wording qualifying the distortion figure confirms measurement under such test conditions. The absence of such a reference dose not necessarily imply that distortion has *not* been measured to this standard.

WOW AND FLUTTER

Both 'wow' and 'flutter' are forms of sound pitch distortion caused by momentary *speed changes* in playback (i.e., on record or tape decks). They produce an unsteadiness in sustained notes—the 'wow' in the lower frequency range and 'flutter' in the higher frequencies.

Wow and flutter is quoted as a single percentage figure, which should be substantially under 1 percent for its effect to be undetectable. Unfortunately, two different standards of measurement are used—peak (or DIN) measurement, and RMS measurement. The former designates the maximum rate of wow and flutter present and the latter a mathematical average of the variations (root mear square). As a result, a peak measurement figure can be 2 to 3 times higher than the RMS figure for the same degree of wow and flutter In other words, quoted figures for 0.3 percent (peak) by one manufacturer and 0.1 percent (RMS) by another could represent identical performance.

RUMBLE

'Rumble' is mechanical noise generated by drive motors in decks. Bearings carrying rotating spindles can also be a further source of 'rumble'. The effect is that of producing a low pitched sound which can be obtrusive during quiet passages of music.

Ideally rumble should be reduced—or filtered out by electronic circuits—so that its level is below that of the limit of audibility. All top class decks are designed so that rumble is suitably eliminated. Decibel figures may be quoted for rumble in the case of record players, when *minimum* requirements for Hi-Fi are:

- 35 dB (unweighted)
- 55 dB (weighted)

Correctly such figures should be measured at a reference frequency of 1,000 Hz and speed of 10 cm/sec.

SPEED (OR MORE CORRECTLY VELOCITY)

Speed—or more correctly, *velocity*—is a measure of the rate of deflection of a stylus traversing a gramophone record groove—not the speed at which the stylus is moving *along* the groove. The

296

response of the stylus, in generating an electrical signal, is dependent on its velocity. It travels at a constant *forward* speed *along* the groove with low *velocity* over undulations representing low recorded sound and higher *velocity* over undulations representing higher recorded sound. Thus *velocity* is the thing which detects differences in sound levels over similar lengths of groove. In other words, the strength of the electrical *signal* generated by a pickup is dependent on its velocity and so velocity, in the units cm/sec, appears in a pick-up performance specification for *output.* The signal strength is measured in *millivolts* (abbreviated mV) and stated for a specific *velocity* in centimeters per second (cm/sec). Output signal strength figures can *only* be compared on this basis, or reduced to millivolts per cm/sec (mV/cm/sec). For example, a simple output figure of, say, 5 mV could mean 5 mV measured at a velocity of 1 cm/sec; or 5 cm/sec; or some other value. In this case an output of 5 mV for 1 cm/sec; or some other value. In this case an output of 5 mV for 1 cm/sec is *five times greater* than 5 mV generated at 5 cm/sec. So to be meaningful, output figures must give both mV and velocity at which this applies. Common standards for velocity are 1 cm/sec and 5 cm/sec.

CROSSTALK

'Crosstalk' is a leakage of a proportion of the signal from one 'input' or 'output' into the channel of a second 'input' or 'output'. It is particularly significant in stereo as excessive crosstalk can destroy an intelligible stereo reproduction.

Crosstalk performance is usually specified in decibels measured at a frequency of 1,000 Hz (or over a frequency range). In practice the crosstalk figure will inevitably be worse at the lower and higher ends of the frequency range, so a crosstalk figure quoted for a frequency range will be worse than one for the single (standard) frequency. An acceptable crosstalk figure is − 20 dB at 1,000 Hz (the lower the figure—i.e., the larger the negative quantity—the better).

POWER! WHAT IT MEANS

Electrical power is measured in *watts*. Thus, for example, the output power of an *amplifier* is quoted as so-many watts, the greater the number of watts the more powerful the amplifier.

Speakers have to be selected to suit that power. Thus if amplifier output is quoted as 2 × 20 watts, then each speaker should be rated as having a power input (or just power) of 20 watts, or more.

Speaker power rating is *not* a simple measurement of how loud they will sound—merely their ability to *accept* a certain level of electrical power *input*.

There is no harm in using speaker powers in excess of the amplifier power—e.g., 30 watt speakers with a 20 watt amplifier. But it can be damaging to *overpower* speakers—e.g., using 10 watt speakers with a 20 watt amplifier operated at full volume.

This is simple and easy to understand—except for one thing. There are several ways of *measuring* power. Also in the case of speakers, power levels received from the amplifier can vary enormously during a program, as well as depending on the setting of the volume control. So power ratings are rather nominal in this case.

As to the different ways of measuring amplifier output power, this is also complicated by the fact that the signal strength is varying all the time. It is not being supplied in the form of a steady current and steady voltage. If it were, using meters to measure the current (in ohms) and voltage (in volts) and multiplying together would give the power in watts. This power has to be determined as some sort of 'average' figure, which can be done in several different ways, all yielding different values for what is essentially the same power!

Only two of these methods need be considered. The most scientific is 'RMS' power or the figure obtained by the 'root mean square' average of the amplifier output power when handling continuous steady tones. The other is based on the power level of a more 'musical' signal and is generally referred to as *music power*—i.e., a measure of the power level when handling a typical musical program. It is also known as IHFM power (or rating), because this method of power measurement was originated by the (American) Institute of High Fidelity Manufacturers.

You will normally find either—or both—power figures quoted in amplifier specifications. It is more important to know which figure is being quoted as for the *same actual power,* the figure for *music power* can be 30-100 percent *higher* than that for rms power (there is no simple conversion factor between the two).

HOW MUCH POWER

Here the answer is not quite so simple, but it really depends on two things:

(i) How loud you want your music to be.

(ii) How large your room is.

Despite what has been said about *speaker* power not being a measure of loudness, *amplifier* power is. For the same speakers (or those of similar efficiency), the more powerful the amplifier the louder the speakers can be made to play.

It does not really matter if the potential maximum loudness is too great. There is a simple answer to this—turning down the volume control. So on this basis, choose an amplifier power which is greater than you are likely to need, and speakers matched to this power.

The only snag here is that until having heard amplifiers of different power playing in any particular room it is difficult to know what the actual loudness is likely to be. The following, however, is a good general guide, the power levels referred to being *rms power* (you can virtually double these figures if working in terms of *music power*).

5 watt amplifier—generally adequate for a smaller room and the *maximum* required for in-car entertainment.

10 watt amplifier—probably about right for an average size room in a modern house, but 'pop' enthusiasts could find it too quiet.

15 watt amplifier—probably about right for a larger living room, or for 'pop' in a smaller room.

20 watt amplifier—Could be regarded as too powerful for domestic listening—but there is always the volume control!

But before making a final decision, read Chapter 13 on loudspeakers.

Chapter 35

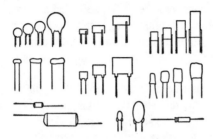

Electronic Control of
Radio and Audio Equipment

The ready availability of MSI and LSI circuits at relatively low cost has made it possible to improve the performance and control facilities of numerous types of domestic equipment, notably radio and audio equipment. Some of the more significant improvements offered by the use of complex integrated circuits are:

(i) Digital frequency and channel indicators which are much more accurate than conventional dials or tuning meters. Elimination of tuning dial also eases styling restrictions.

(ii) *Dc* control of analog and switching functions. This eliminates spurious coupling problems, thereby simplifying component layout and lifting styling restrictions. This type of control also facilitates remote control of all the functions in electronic entertainment equipment.

(iii) Tuning-voltage memory for pre-selecting chosen programs.

(iv) Improved tuning stability in AM and FM radios by using phase-locked loop frequency synthesizers.

(v) Automatic tuning with built-in criteria to automatically select stereo broadcasts, traffic information services, transmissions with a minimum strength etc.

FREQUENCY MEASUREMENT AND DISPLAY FOR RADIOS

The following description by Philips/Elcoma describes their system designed to measure the frequency to which an AM/FM radio is tuned and to provide a digital display of either the tuned-frequency or the associated FM channel number.

The system counts the number of receiver local-oscillator cycles that occur during a pre-determined period, subtracts a pro-grammable intermediate frequency from the count and feeds the resultant frequency to a display decoder/driver. The output from the decoder is displayed in the form of 4½ seven-segment digits or, in the case of FM, there is an option of displaying the associated two-digit channel number.

Since the system requires only two integrated circuits and a few passive peripheral components in addition to the display, it occupies very little printed-wiring board area. Furthermore, it can be programmed to operate on long-wave, medium-wave, short-wave and vhf in association with a wide range of intermediate frequencies. The system can therefore be very easily added to the majority of existing mains-driven radio receiver designs. The following are outstanding features of the system:

(i) Mains zero-crossing switching to reduce interference.

(ii) Multiple sampling to stabilize display during short-term local-oscillator drift.

(iii) Suitable for use with most types of seven-segment display.

(iv) Suitable for use with a wide range of intermediate frequencies, with AM, 449 kHz to 472 kHz, on FM, 10, 6 MHz to 10,775 MHz.

(v) Compact circuitry with few peripheral components.

(vi) Facilities for 'freezing', testing and blanking the display.

(vii) Wide frequency range and high resolution display:
vhf channel number: +02 to +64 to within 0,1 MHz
vhf frequency: up to 109,35 MHz to within 0,05 MHz
SW frequency: up to 19,995 kHz to within 5 kHz
MW/LW frequency: up to 1,999 kHz to within 1 kHz

(viii) flicker suppression.

(ix) high input sensitivity allows direct drive from radio local oscillators.

(x) If an LED display is used, the system only requires a single 8 V ac supply.

BRIEF DESCRIPTION

The principle of the frequency measuring and display system is illustrated by the block diagram in Fig. 35-1. The main components

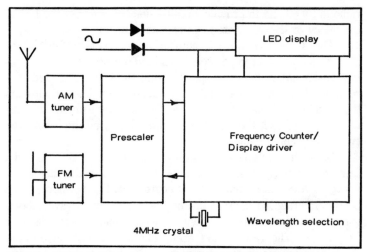

Fig. 35-1. Block diagram of frequency measuring and display system.

of the system are an integrated programmable pre-scaler type
SAA1058, an integrated frequency counter/display driver type
SAA1070, a 4 MHz quartz crystal and a 4½-digit seven-segment
display.

The pre-scaler contains a pre-amplifier with inputs for AM and
FM local-oscillator signals, a frequency divider and two output
amplifiers for driving ECL, TTL or MOS circuits. The high input
sensitivity of 5 mV for AM and 10 mV for FM oscillators, thereby
eliminating the need for an amplifier interface.

The pre-scaler divides the local oscillator frequency by 32 and
provides a square-wave output to the integrated frequency
counter/display driver. The frequency is then measured by counting the number of negative-going transitions that occur during a
period defined by a crystal-controlled pulse generator. The same
pulse generator presets the pre-scaler at the start of each counting
period to minimize indecisive counting and thereby reduce display
flicker. Nine of the fifteen output pins of the SAA1070 are also used
to program the system for use with the appropriate intermediate
frequencies for AM and FM. Selection of waveband and display
mode (frequency/channel number/display test/display blanking) is
effected by connecting one of four pins of the SAA1070 to the supply
return line. Applying a mains-frequency half sine-wave to the DUP
input of the circuit results in the display being driven in the duplex
mode and minimizes interference radiation by ensuring that
switching of the display segments can only occur during the mains

waveform zero crossing. The circuit is capable of directly driving the segments of a LED display so that discrete driver transistors are not required.

PROGRAMMABLE PRE-SCALER

The SAA1058 is a multi-stage divider with an externally selectable division ratio of 32:1 or 33:1. This facility allows the circuit to also be used in frequency synthesizer systems such as our microcomputer-controlled phase-locked-loop tuning system which requires the use of both ratios. In the system being described, only the 32:1 division ratio is used.

Figure 35-2 is a simplified functional block diagram of the pre-scaler. The preamplifier ensures a high input sensitivity so that signals from the local-oscillators in the radio can be connected to the system via a passive coupling network. The AM and FM local-oscillators can be simultaneously connected to the symmetrical inputs of the pre-amplifier without an input filter or a switch provided that the unused oscillator is switched off in the radio. The pre-scaler incorporates two symmetrical output preamplifiers, each of which drives a different kind of complementary output stage. The outputs are thus made compatible with ECL, TTL or MOS inputs and can drive circuits that react to either the falling or rising edges of the drive signal. Each of the four internal functional blocks has its own supply pin; pin 3 for V_{CC1} to the input pre-amplifier, pin 14 for V_{CC2} to the synchronizing stage, pin 12 for V_{CC3} to the divider, and pin 10 for V_{CC4} to the output stages.

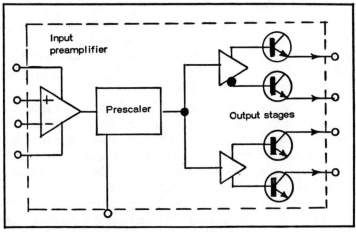

Fig. 35-2. Simplified functional diagram of the pre-scaler.

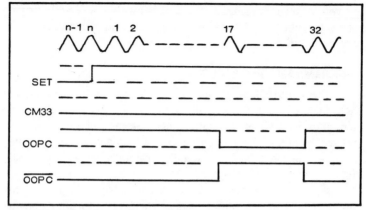

Fig. 35-3. Timing diagram for pre-scaler.

Figure 35-3 is the timing diagram for the pre-scaler. When input CM33 is set low, the division ratio is set to 32:1. As long as the SET input is high, a transition occurs at each of the outputs after every 16 input cycles. To ensure that the least significant digit of the displayed frequency is least likely to flicker due to phase shift between the counting period and the pulses being counted, the first output transition must ideally occur coincident with the 17th input frequency period after the start of each counting period. This synchronization is achieved by applying a low-to-high transition, representing the start of the counting period, to the SET input of the pre-scaler.

FREQUENCY COUNTER/DISPLAY DRIVER

The integrated frequency counter/display driver type SAA1070 is specifically designed for the control of a duplex seven-segment display of the frequency or FM channel number to which an AM/FM radio is tuned. A simplified block diagram is given in Fig. 35-4. The main features of the circuit are:

(i) Output stages capable of directly driving a 4½ digit LED frequency/channel number display with LED lamp indication of kHz and MHz. To minimize the number of output pins, and to reduce radiated interference, the segment drive outputs are driven in duplex mode in synchronism with the zero crossing of the mains-derived supply connected to the display anodes.

(ii) 18-bit frequency counter which can be pre-set to compensate for a wide range of externally programmed intermediate frequencies for the vhf, shortwave and medium/long-wave bands.

304

(iii) 16-bit comparator for comparing the measured and displayed frequencies and updating a 16-bit display register if there is a discrepancy for three successive measuring periods. This technique prevents display flicker due to short-term drifts of the local oscillators in the radio.

(iv) Seven-segment decoder and multiplexer for driving the output stages.

(v) Internal timing pulse generator controlled by a 4 MHz quartz crystal.

(vi) A display 'freeze' facility which allows the last frequency to be measured is to be continuously displayed.

(vii) Display test and blanking facilities.

The functions of the frequency counter/display driver will be explained with the aid of the simplified block diagram given in Fig. 35-4.

After the radio local-oscillator signal has been divided by 32 in the prescaler, it is fed to input FIN (pin 12). The counter then determines how many pulses have occurred during a measuring period defined by a pulse generator under the control of a 4 MHz quartz crystal connected between pins 17 and 18. Since the counter has been preset with a number corresponding to the i-f of the radio, the pulse count is proportional to the frequency to which the radio is

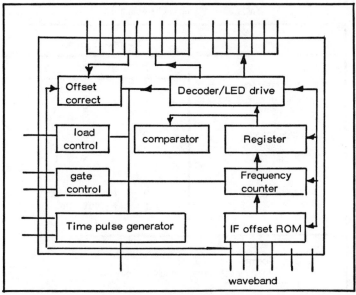

Fig. 35-4. Simplified block diagram of frequency counter/display driver.

tuned. The result of the count is compared with the displayed frequency and, if there is a difference for three successive measuring periods, the display register is updated on occurrence of the next synchronizing pulse (DUP) at pin 16.

Before the SAA1070 is ready for operation, it must be set up for use in accordance with the operating conditions of the radio and of the display as described under the next two headings.

WAVEBAND SELECTION AND I-F OFFSET

Pins 8, 9, 10, and 11 must be programmed in accordance with Table 35-1 to suit the waveband in use and, in the case of vhf, to determine whether the channel number or the tuned-frequency of the radio will be displayed. This program also determines the display resolution (Table 35-2) by presetting the 4 MHz crystal-controlled timing pulse generator to determine a measuring period of the following duration:

vhf and channel no: 256 μs (1,024 cycles of 4 MHz)
short-wave: 2,56 ms (10,240 cycles of 4 MHz)
medium/long-wave: 3,2 ms (12,800 cycles of 4 MHz)

The waveband program also addresses the i-f offset ROM to preset the frequency counter to compensate for an intermediate

Table 35-1. Waveband Selection Inputs.

Selection	SAA 1070 pin no.			
	8	9	10	11
Uhf frequency	1	1	1	0
Uhf channel	1	1	0	X
Short-wave	1	0	X	1
Medium/long	0	1	X	1
wave	0	1	0	0
Display test	X	0	X	0
	0	0	X	1
Display blank	0	1	1	0
	1	1	1	1

0 = Common return or pin 1
1 = Not connected or connected to pin 14
X = 0 or 1

306

Table 35-2. Display Resolution.

Waveband	Resolution	Equivalent number of transitions at input of SAA 1070
Uhf frequency	0,05 MHz	4
Uhf channel*	0,1 MHz	8
Short-wave	5 kHz	5
Long/medium wave	1 kHz	1

one channel = 300 kHz
channel 02 = 87,6 MHz

frequency of 10,7 MHz for vhf and 460 kHz for short/medium/long-waves. If the required intermediate frequencies differ from these values, the i-f offset pulse generator must be preset by programming input/output pins 20 to 28 as shown in Tables 35-3 and 35-4. These pins change function from segment drive outputs to i-f offset inputs for a short period coincident with the negative-going edge of the synchronizing pulse applied to the DUP input at pin 16.

SYNCHRONIZING PULSE

The start of the entire frequency measuring and display switching sequence is synchronized with the negative-going edges of pulses applied to the DUP input at pin 16. If mains-derived half-sinewaves are used to power the LED duplex display, it is convenient to use one of them as a synchronizing pulse. This will ensure that radiated interference is reduced by effecting the decoded segment output switching at the instant when the mains waveform crosses zero.

THE SIGNAL PROCESSING SEQUENCE

To clarify the signal processing sequence of the frequency measuring and display system, each of its twenty intervals will be described in turn with the aid of Fig. 35-5. Each interval lasts for a period determined by a number of 4 MHz cycles as demanded by the program at the waveband address inputs of the SAA1070. The timing sequence is started by the negative-going flank of the mains-derived half-sinewave synchronizing pulse (DUP) applied to pin 16 of the SAA1070.

Table 35-3. Offset Correction Inputs for AM.

| SAA 1070 Input-pin | | | | | SW | M/LW |
21	22	25	26	28	kHz	kHz
0	0	0	0	0	460,00	460
0	0	0	1	0	448,75	449
1	0	0	1	0	450,00	450
0	1	0	1	0	451,25	451
1	1	0	1	0	452,50	452
0	0	1	1	0	453,75	453
1	0	1	1	0	455,00	454
0	1	1	1	0	456,25	455
1	1	1	1	0	457,50	456
0	0	0	0	1	456,25	457
1	0	0	0	1	457,50	458
0	1	0	0	1	458,75	459
1	1	0	0	1	460,00	460
0	0	1	0	1	461,25	461
1	0	1	0	1	462,50	462
0	1	1	0	1	463,75	463
1	1	1	0	1	465,00	464
0	0	0	1	1	463,75	465
1	0	0	1	1	465,00	466
0	1	0	1	1	466,25	467
1	1	0	1	1	467,50	468
0	0	1	1	1	468,75	469
1	0	1	1	1	470,00	470
0	1	1	1	1	471,25	471
1	1	1	1	1	472,50	472

0 = no connection
1 = 22kΩ resistor connected to 2,5V

Interval 1

Pins 20 to 28 of the frequency counter/display driver are switched over to function as i-f offset correction address inputs and the required program is stored in the i-f offset correction circuit.

Interval 2

The i-f offset ROM contents addressed by the waveband selection circuit is loaded into the frequency counter. At the end of the interval, pins 20 to 28 of the frequency counter/display driver are switched back to their function as display segment drive outputs.

Interval 3

The program stored in the i-f offset correction circuit causes the required number of pulses to be loaded into the frequency counter.

Table 35-4. I-F Offset Correction Inputs for FM.

SAA 1070 input pin				i.f. (MHz)
20	23	24	27	
0	0	0	0	10,70
1	0	0	0	10,60
0	1	0	0	10,6125
1	1	0	0	10,625
0	0	1	0	10,6375
1	0	1	0	10,65
0	1	1	0	10,6625
1	1	1	0	10,675
0	0	0	1	10,6875
1	0	0	1	10,70
0	1	0	1	10,7125
1	1	0	1	10,725
0	0	1	1	10,7375
1	0	1	1	10,75
0	1	1	1	10,7625
1	1	1	1	10,775

0 = no connection
1 = 22 kΩ resistor connector to 2,5V

Interval 4

The timing pulse generator is programmed by the waveband selection circuit to generate a measuring period of the appropriate duration.

Intervals 5 to 14

At the end of interval 4, the GATE output from the SAA1070 goes high, thereby activating the SET input to the pre-scaler and initiating the output pulse train. These pulses enter the SAA1070 at the FIN input at pin 12 from where they pass via a measuring-period gate to be loaded into the frequency counter. At the end of interval 14 the timing pulse generator concludes the pulse counting by disabling the measuring-period gate.

Interval 15

This interval is free.

Interval 16

During this interval, the 16 most-significant bits from the 18-bit frequency counter are compared with the contents of the 16-bit display register. If there is a discrepancy, a 2-bit counter in the comparator is incremented by one. If the contents of the 2-bit counter are then less than 3, the sequence proceeds to interval 17. If incrementing the 2-bit counter causes its contents to reach 3 (contents of frequency counter and display register different for three consecutive measuring periods), the timing pulse generator is stopped before the sequence proceeds to interval 17. If the contents of the frequency counter and the display register are the same, the 2-bit counter is reset before the sequence proceeds to interval 17.

Interval 17

The system waits for the negative-going flank of the next half-sinewave to appear at the DUP synchronizing input. If the contents of the 2-bit counter are less than 3, the process then recycles to interval 1. If the 2-bit counter has reached a count of 3, the timing pulse generator is re-started and the sequence proceeds to interval 18.

Interval 18

The contents of the frequency counter are loaded into the display register and the display is thereby changed to the new value.

Interval 19

This interval is free.

Interval 20

The timing pulse generator is stopped and the system waits for the negative-going flank of the next half-sinewave to appear at the DUP synchronizing input of the SAA1070. The timing pulse generator is then restarted and the sequence recycles to interval 1.

DISPLAY OPTIONS

The sequence of events just described only remains valid as long as the DISP input at pin 19 of the SAA1070 remains discon-

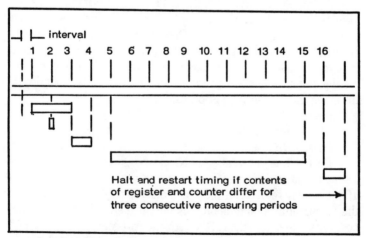

Fig. 35-5. Signal processing sequence.

nected. The DISP input can be used to change the mode of operation as follows.

DISP INPUT CONNECTED TO COMMON RAIL

If the DISP input is connected to the common rail, the sequence is halted at the end of the sequence because the timing pulse generator is inhibited so that the GATE output from the SAA1070 becomes continuously low, thereby holding the pre-scaler in the SET state and inhibiting its output pulses. The displayed value is therefore 'frozen' at the last frequency to be measured.

DISP INPUT CONNECTED TO SUPPLY VOLTAGE

If the DISP input is connected to the same supply voltage as pin 14 of the SAA1070, the displayed frequency is changed whenever there is a difference between the contents of the counter and the display register during interval 16, regardless of the contents of the 2-bit counter. In this mode of operation, the flicker reduction facility is lost but the response time of the system is reduced.

THE FREQUENCY MEASUREMENT AND DISPLAY SYSTEM

In its simplest form, the system is intended for LED display of frequency or FM channel number in mains-powered radios. Only a single 8 V 50 Hz supply is then required. The system can however also be adapted for use with dc supply of 11 V to 16 V (batteries) if an additional duplex pulse generator and a voltage stabilizer are used. Furthermore, other types of display can be accommodated if a slightly more complex power supply and segment drive circuit are used.

Chapter 36

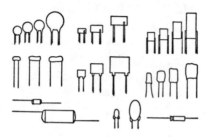

Rhythm Generators

A rhythm generator is defined as a system or circuit which generates trigger pulses for an oscillator (or oscillators) whose amplified and damped output(s) simulate the sound(s) of the musical instruments in the rhythm section. They are thus basically a counting/timing device with a fixed memory. Counting involves dividing each cycle of the complete rhythm with a number of elemental times or *counter states* the memory then determines whether or not a given instrument should be triggered during those counter states.

The counter states, which constitute the smallest subdivisions of the rhythm can be grouped into bars or measures (usually 1, 2, 3 or 4). Within the complete rhythm, each of these bars can be programmed differently (e.g., the bossa nova).

Each bar, then, consists of n elementary times in which the beats of each instrument will be programmed to occur. In terms of musical notation the length of these beats is described as a fraction of a known reference period (See Fig. 36-1).

When the sum of the beats in any bar comes to 4/4, the rhythm is described as 4/4. Similarly it is possible to have a 3/4 rhythm and so on.

The number of elementary times in the bar fixes the minimum duration of each beat; in other words, the greater the number of elementary times the shorter will be the minimum duration of each beat; in other words, the greater the number of elementary times the shorter will be the minimum length of the beats and the richer the resulting rhythm.

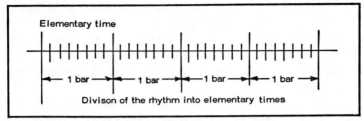

Fig. 36-1. Relationship between elementary times and the bars of a rhythm.

For example, a 4/4 rhythm programmed in 4 bars over 32 elementary times, i.e., 8 per bar, can only use musical beats of length 1, ½, ¼, or ⅛ and not of 1/16, 1/32, 1/64.

If the same rhythm is programmed in 2 bars of 16 elementary times each, musical beats of length 1, ½, ¼, ⅛ and 1/16 can be used, 1/32 and 1/64 being still excluded.

Figure 36-2 shows a block diagram of a typical rhythm generator. The counter must be able to count the number of elementary times corresponding to rhythms of 3/4, 4/4 and 5/4.

This means that the counter must stop and reset to its initial position (to repeat the rhythm) after a certain number of counts which depends on the selected rhythm. Two characteristics of the rhythm determine the count requirement:

(i) the minimum beat length and
(ii) the number of bars in the complete rhythm

For example, if the rhythm is 4/4 with a minimum duration of 1/16 and 2 bars per theme the count = 16 elemental times × 2 bars = 32 counter states. Similarly a 3/4 rhythm with a similar minimum duration and number of bars per theme would give a count

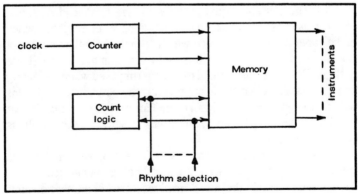

Fig. 36-2. Block diagram of a trigger generation system for the oscillator circuits.

requirement of $16 \times 3/4 \times 2 = 24$ counter states; or a 5/4 rhythm $16 \times 5/4 \times 2 = 40$ counter states.

The memory needs to be of ROM type and must have outputs which reset to zero after each read-out so that the output will always be able to provide the correct trigger edge during the following beat.

Such circuits can be designed and constructed with discrete components, or more simply and conveniently using a specialized IC. In the latter part is commonly produced in TTL, with the memory in MOS to provide the necessary number of bits in a convenient package size. A certain difficulty arises in that ROM outputs do not automatically reset to zero, so an external clock is usually necessary for reset in the case of ICs. Using discrete components, both memory and counter decoder would normally be chosen in the form of a diode matrix which could lead to a very large number of diodes being required.

DESIGN REQUIREMENTS

Design requirements for an ideal rhythm generator are:

(i) The entire system described above would be contained in a single device thereby achieving maximum reliability in the minimum space. The labor required for assembly is also minimized.

(ii) The counter should have the highest count possible. For a rhythm containing a fixed number of bars this means that the rhythm can be subdivided into shorter beats and will consequently be musically more interesting. Similarly for a given number of elementary times the rhythm can be made up of a greater number of bars (possibly different) resulting, once again, in a more interesting musical effect.

(iii) The system should provide a large number of rhythms. Here it is necessary to make a distinction between rhythms which can be superimposed and those which cannot, since the concept of superimposition is closely linked to the number of available rhythms. Two rhythms are said to be superimposed when, selecting both simultaneously at the system input, the output commands for each instrument correspond to a combination of the commands that would have been produced by the rhythms selected separately as shown in Fig. 36-3.

Technically, rhythms can only be superimposed if they are selected by means of separate lines and not a coding technique.

Superimposition, therefore, involves a greater number of input pins (one for each rhythm), but does not call for a very high number

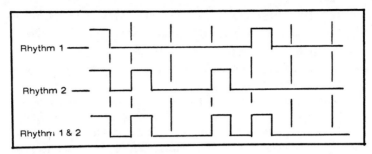

Fig. 36-3. Combination of two rhythms.

of rhythms since the organist can choose any combination of those available.

In general, 12 is a sufficient number of rhythms if they can be superimposed, but there have to be more if superimposition is not possible, usually 15 or 16.

(iv) The system must have a large number of outputs (instruments). The number of instruments programmed for each rhythm will generally vary between 3 and 6. Eight represents a maximum number which is rarely used.

(v) The system must be programmable in any time, 3/4, 4/4, 5/4, 6/8. It must therefore be possible to intervene at the mask programming stage, on the reset of the elementary time counter for each rhythm.

(vi) The system must produce no spikes on the memory outputs caused by momentary coincidence of two successive decoded counter states. This can create undesired triggering of the instrument oscillators.

(vii) The system must provide the possibility of externally resetting the elementary time counter, so that it restarts from the first elementary time of the first beat. This enables KEY or TOUCH operation in which the rhythm generator remains reset until at least one key is played.

(viii) The system should supply a down-beat output signal corresponding to the first elementary time of the first beat of each rhythm. This signal allows synchronization between the organist and the device's internal counter.

(ix) The system must be related with a static form of logic designed for the low frequency operation of a rhythm generator (20 Hz).

(x) The system must be input compatible with TTL and DTL level signals so that it can be interfaced with an oscillator realized with such devices.

(xi) The system must have low dissipation (150 to 300 mW).
(xii) The system must have a single standardized supply.

Points (ii), (iii) and (iv) together create a single requirement, namely that the system must have a maximum memory capacity in terms of the number of bits.

The maximum number of bits is limited by the die-size of the device which in turn is determined by the cost of the device itself.

Once the memory capacity has been established by economic factors, it follows that a compromise between the number of rhythms, the number of instruments and the number of elementary times will be made. An effective solution is:

> a maximum of 32 elementary times
> 8 instruments
> 15 rhythms

with a memory capacity of 3840 bits ($32 \times 8 \times 15$).

HARDWARE

Turning to specific hardware requirements, one would normally choose an IC design for the job rather than consider discrete component circuitry, for reasons already mentioned above. There are many such devices available, produced by different manufacturers. For the sake of description we will consider true specific types—the SAS-ATES M253 and M252.

The M253 features:
— 12 rhythms which can be superimposed
— 8 instruments for 3/4 or 4/4 time, or 7 instruments
   for any time
— a maximum of 32 elementary times
— external reset
— down-beat output
— internal anti-spike circuit
— single supply
— minimum dissipation (typically 100 mW)
— can be direct interfacing (clock) with TTL, DTL
— pin to pin compatible with the M250
— fully direct-coupling
— 24-pin plastic or ceramic package

The M252 features:
— 15 rhythms which cannot be superimposed
— 8 instruments for 3/4 or 4/4 time, or 7 instruments for any
   time

—a maximum of 32 elementary times
—external reset
—down-beat output
—internal anti-spike circuit
—single supply
—minimum dissipation (typically 100 mW)
—direct interfacing (input) with TTL and DTL
—fully direct-coupling
—16-pin plastic or ceramic package

Both of these devices are derived from the same chip which, during the processing stages, is provided with the memory pattern specified by the customer and the ancillary functions that distinguish the particular system.

USING THE M253

A block diagram of the M253 is shown in Fig. 36-4. The phase generator uses the incoming clock signal to produce the 2 non-overlapping phases at regenerated levels which are required for driving the following divider.

This divider has to create a reset signal for the return to zero of the outputs. The width of this pulse is independent of the duty cycle of the incoming clock. The divider's outputs also serve as timing signals for the first stage of the 5-stage counter, which uses master-slave flip-flops. The counter states are decoded to drive the rows of the memory matrix. The columns of the matrix are divided into 12 groups of 8, representing the 12 rhythms and the 8 instruments.

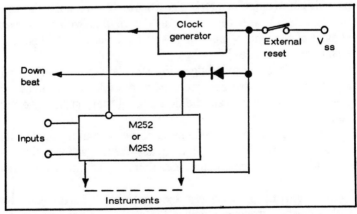

Fig. 36-4. Restarting after an arbitrary number of elementary times.

One particular state, the 24th, is decoded, logically combined with rhythms in 3/4 time and is used as the counter's internal reset for rhythms programmed in this time. This device is therefore suitable for programming any rhythm in 4/4 time over 32 elementary times or in 3/4 time over 24 elementary times. This means that when a rhythm is programmed over a single bar the intervals can be as short as 1/32 allowing great musical flexibility.

The counter can also be reset by an external signal which, when driven directly by an output of the M253 itself, sacrificing one instrument, can be used to reset the counter to any position for times other than 3/4 or 4/4. If, for example, we want to reset at the state n for a rhythm x, a beat must be programmed at the elementary time $n + 1$ at the output of the rhythm to be used as reset (Fig. 36-4).

The down-beat impulse lasts only for 2 - 3 μs so if it is to be used to drive a lamp it must first be stretched and buffered.

This output, connected at the input of the external reset, immediately zeroes the counter and therefore causes the disappearance of the reset signal (other than the $n + 1$ beat there should be no program on the output used as the reset).

The columns of the matrix are enabled singly or in groups (the rhythms can be superimposed) via the buffer according to the rhythm or rhythms selected.

The presence of one rhythm, therefore, does not exclude the possibility of another rhythm being selected contemporarily, and the result on the output of each instrument for each rhythm is the sum of the beats of the single rhythms as already shown.

One particular case is when the rhythms are selected contemporarily with a different matrix e.g., a 3/4 or 4/4 rhythm. In this case the count cycle will correspond to the rhythm with the lowest number of elementary times (in the example the cycle will be of 24 elementary times).

The delayed, decoded signal from the 24th state (3/4 rhythms) and the 32nd (4/4 rhythms) are used as down-beat signals, i.e., as starting signals to indicate the first beat of the first bar (see Fig. 36-5).

This signal, whose usefulness will be seen in the application section, was brought out to a pin already used for an input signal—the external reset signal—since no supplementary pin was available in the package used.

In reality the presence of an external reset signal is compatible with a down-beat signal although the reverse is not true since a

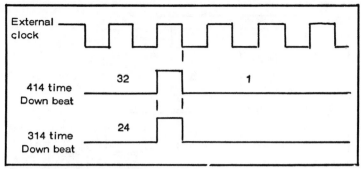

Fig. 36-5. Down-beat timing and duration.

down-beat signal must not have the effect of an external reset. This can be achieved by using a diode to separate the two signals as shown in Fig. 36-6.

In the case of rhythms other than 3/4 or 4/4, the pulse present at the output connected to the external reset, can be used to trigger a monostable circuit whose output will be the down-beat signal. When no rhythm is selected, the down-beat signal is present and the counter counts to 32.

USING THE M252

A block diagram of the M252 is shown in Fig. 36-7. Here the phase generator, the counter, the matrix, the output and reset logic, and the 24th state decoder for the reset in 3/4 time, operate in the same way as in the M253. The difference is in the rhythm command inputs, which are in binary logic using the code shown in Table 36-1.

Given the fact that it is impossible to select two different codes at the same time, it follows that it will be impossible to superimpose

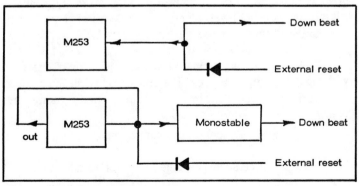

Fig. 36-6. Using the down-beat signal.

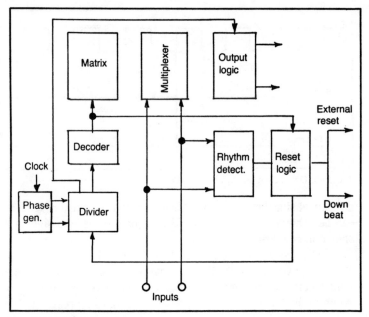

Fig. 36-7. M252 block diagram.

these rhythms. One code word has been used to indicate << no rhythm selected>>. In this state, there are no instrument output signals, the down-beat signal is present and the counter counts to 32.

TIMING WAVEFORMS

With the dynamic characteristics available it should be noted that a duty cycle of 50 percent is not required for the clock signal. The width of the <<mark>> of the clock waveform need only be as great as the width of the down-beat impulse internally generated. All the supplies and levels shown in the electrical characteristics are expressed as function of V_{SS}. Since $V_{SS} = 17$ V; $V_{GG} = -12$ V; $V_{SS} = +5$ V, and so on as long as $V_{SS} - V_{GG} = 17 \pm 1$ V.

This makes it very simple to solve the problem of interfacing with input and output devices. The builder, however, must always respect the limits imposed by the absolute maximum ratings given below.

Absolute maximum ratings

V^*_{GG} Source supply voltage	-20 to 0.3 V
V_1 Input voltage	-20 to 0.3 V

I_o	Output current (at any pin)		3	mA
T_{stg}	Storage temperature	-65 to 150		°C

* This voltage is with respect to V_{SS} pin voltage

These voltages, temperatures or currents are values which must never be exceeded, not even momentarily, since the device can be permanently damaged should this occur. It is of particular importance that the builder keeps a check on the positive overshoot at all the pins with respect to V_{SS}.

If the positive overshoot, which, on the oscilloscope will always appear limited to one V_{BE} when measured on an in-circuit device, exceeds the values quoted in the absolute maximum ratings it causes a parasitic which discharges the surrounding negative nodes, causing incorrect circuit operation. More seriously, a fixed positive level more than 300 mV above V_{SS} will probably damage the circuit.

Encoders

Various encoders can be used for selecting the rhythms—e.g., diode matrix, TTL encoder, COSMOS encoder. A typical COSMOS

Table 36-1. M 252 Rhythm Selection Code (positive logic).

Rhythm	Code			
	IN 8	IN 4	IN2	IN1
1	1	1	1	0
2	1	1	0	1
3	1	1	0	0
4	1	0	1	1
5	1	0	1	0
6	1	0	0	1
7	1	0	0	0
8	0	1	1	1
9	0	1	1	0
10	0	1	0	1
11	0	1	0	0
12	0	0	1	1
13	0	0	1	0
14	0	0	0	1
15	0	0	0	0
No selected rhythm	1	1	1	1

encoder, however, may need modifying for single-pole switches (e.g., inserting resistors in each switch lead).

Clock Generator

Choice here would be TTL or DTL, HLL, or COSMOS, or even discrete components.

Application to an Electronic Organ

In order that the rhythm section may be inserted in the organ, a signal must be available which indicates whether one or more keys on the organ keyboard have been played.

This signal, which we shall call the key played, starts the rhythm section. When a key is played the rhythm section can be arranged to start at the beginning of the bar (touch or key operation), i.e., the playing of a key removes the reset from the clock and from the M252 or M253.

Alternatively it can be arranged to start at any point in the bar (continuous or silent operation) i.e., the rhythm generator runs continuously, but its output is enabled by the <<key played >> signal. In continuous operation, therefore, the down-beat indicator is indispensable since it allows the first key to be played when the bar begins.

A third method (continuous free running) allows the unit to operate without playing any of the keys. This is done simply by selecting a rhythm on the push button array of the rhythm section. Neither the touch key nor the continuous silent key must be on when this method is used.

Figure 36-8 is an illustration of the insertion of the electronic rhythm section into an organ. The two parts within dashed lines are details of the rhythm section, of interest for the connections to the keyboard of the organ.

The electronic rhythm section described here was realized with the M252AA, programmed with 15 different rhythms, in such a way that each rhythm can use up to a maximum of 8 of the 9 instruments available.

The 15 rhythms programmed are the Waltz, Jazz Waltz, Tango March, Swing, Foxtrot, Slow Rock, Pop Rock, Shuffle, Mambo, Beguine, Cha Cha, Bajon, Samba and Bossa Nova.

These rhythms can be brought in one at a time by means of the keyboard. The instruments available are the bass drum, snare drum, claves, high bongo, low bongo, conga drum, long cymbals, short cymbals, and maracas.

322

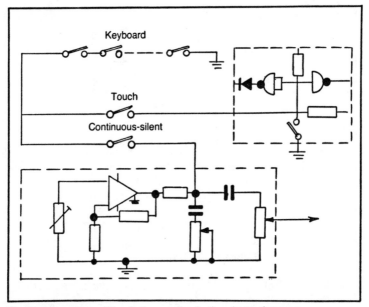

Fig. 36-8. Insertion of the rhythm section in the electronic organ.

Sound Generators

Sound generators should be designed to reproduce as faithfully as possible the sounds made by percussion instruments. They can be divided into two broad groups, namely, sounds consisting of damped, sinusoidal waves, like drums, and those consisting of damped white noise, like cymbals. In the first category we can include the bass drum, high bongo, low bongo, conga drum and the claves, for which the basic circuit is as shown in Fig. 36-9.

This circuit is a simple double-T oscillator with active COS-MOS element kept slightly below the point of oscillation by P1. To obtain the effects of different instruments you only have to select the right values for the capacitors C1 and C2. In this way the frequency of the instrument required can be obtained. The potentiometer P1 also regulates the length of the damping, so that longer or shorter sounds can be obtained.

Since the M252 produces a square wave, the differentiator RC must be introduced so that a fairly short pulse arrives at the oscillator, which should not interfere with the damping of the oscillation but should be sufficient to activate the oscillator itself. The resistor R1 keeps the input at earth in the absence of a command, otherwise it would remain floating since the outputs of the M252AA are open-drain types.

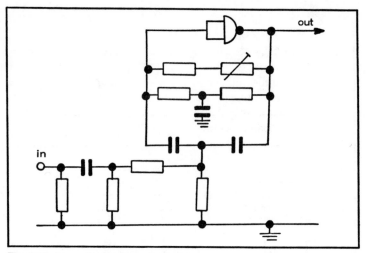

Fig. 36-9. Sinusoidal instrument simulator.

In the second category we find the long cymbals, short cymbals, and maracas, for which the basic circuit is of the kind shown in Fig. 36-10. The white noise produced by the zener effect of the base-emitter junction of a transistor is applied at the base of Q2. During the discharge of C2, therefore, transistor Q2 can amplify this noise. The level of amplification, however, will follow the discharge curve of C2 and therefore a damping effect of variable length will be obtained according to the values of C2 and R2.

The inductor L and the capacitor at the collector of Q2 allow partially selective amplification to be obtained so that some har-

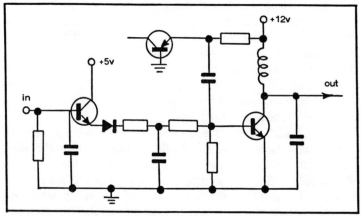

Fig. 36-10. White-noise instrument simulator.

monics can be boosted and an effect more similar to the instrument being simulated can be obtained.

As can be seen from the photographs of the signals, almost all the instruments used in this rhythm section start immediately with maximum amplitude and decrease exponentially. The only exception is the maracas simulator, whose signal increases progressively and then decreases like the others. This effect was achieved by means of the integrator-differentiator circuit which allows controlled amplification of the white noise. The snare drum is obtained by adding a signal of the second type, i.e., a metallic sound, to a drum sound. As can be seen from the photographs and the waveforms shown earlier, each sound starts on the positive edge of the control pulse.

Keyboard

The keyboard must be considered as a separate system having a precise logic function, namely that of encoder. In fact by playing any key the code corresponding to that rhythm is obtained and is applied at the four inputs of the M252AA, $\overline{IN1}$, $\overline{IN2}$, $\overline{IN4}$, and $\overline{IN8}$, thus selecting the rhythm required according to Table 36-2. The keyboard also has the function of connecting the second output of

Table 36-2. M 252 AA Rhythm Selection Code (positive logic).

Rhythm	Code			
	$\overline{IN\ 8}$	$\overline{IN\ 4}$	$\overline{IN\ 2}$	$\overline{IN\ 1}$
Waltz	1	1	1	0
Jazz Waltz	1	1	0	1
Tango	1	1	0	0
March	1	0	1	1
Swing	1	0	1	0
Foxtrot	1	0	0	1
Slow Rock	1	0	0	0
Pop Rock	0	1	1	1
Shuffle	0	1	1	0
Mambo	0	1	0	1
Beguine	0	1	0	0
Cha Cha	0	0	1	1
Bajon	0	0	1	0
Samba	0	0	0	1
Bossa Nova	0	0	0	0
No selected rhythm	1	1	1	1

the M252AA to the snare drum or to the claves according to the rhythm selected.

Rhythm Section IC

A rhythm section with 12 rhythms and 8 instruments can be realized with the IC M253AA. Here 12 different rhythms are programmed, each rhythm being able to drive simultaneously a maximum of 7 out of the 8 instruments available. The 12 rhythms programmed are the Tango, Waltz, Shuffle, March, Slow Rock, Swing, Pop Rock, Rumba, Beguine, Cha Cha, Samba and Bossa Nova. These rhythms can also be combined, two more can be selected contemporarily. The instruments are the same as for the preceding unit with the exception of the conga.

The sound generators, variable clock generator and monostable for the down-beat, are the same as for the preceding rhythm section. The keyboard no longer has the function of encoder, only that of connecting the snare drum or the claves to the third output of the M253AA, according to the rhythm selected

The last important difference between this and the preceding unit lies in the first and third outputs of the M253AA. The first output is not programmed to drive an instrument but controls the alternating bass (basso alternato) of the electronic organ, (Fig. 36-11).

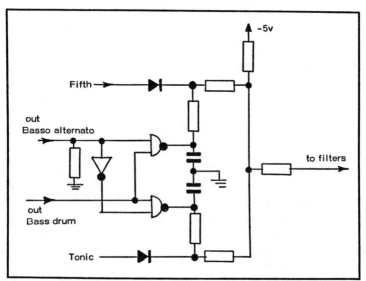

Fig. 36-11. Basso alternator driving circuit.

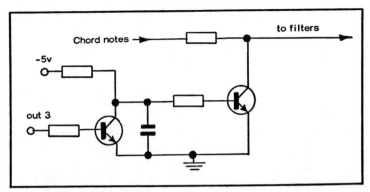

Fig. 36-12. Chord driving circuit.

As this diagram shows the tonic appears when the OUT1 signal is absent and the OUT2 signal is present. The fifth on the other hand comes out when both output 1 and output 2 are present. To conclude, each time that there is a beat of the bass drum (OUT2) a note comes from the basso alternato. The OUT1 serves only to establish which of the two notes will be played.

The third output is programmed to drive one instrument, in this case the snare drum or the claves, and simultaneously to control the output of the chords played on the keyboard of the organ (Fig. 36-12).

Operating the Rhythm Section

By resetting the clock generator to zero instead of to one (positive logic) the bar will begin half a clock-period later than the release of the reset. By leaving the clock generator free, i.e., resetting only the M252 or M253, two things can happen at the release of the reset:

(i) If the clock is in 'O' position the rhythm starts immediately from the beginning of the bar.

(ii) If the clock is at '1' the bar begins as soon as the clock switches over, therefore there is a random delay which varies from about zero to half a clock-period.

For both the M252 and M253 with no reset applied, the clock running and no rhythm selected, the down-beat signal occurs every 32 elementary times or every 64 clock pulses.

Index

Edited by Roland Phelps